Education and Training for the Oil and Gas Industry: Building a Technically Competent Workforce

Education and Training for the Oil and Gas Industry: Building a Technically Competent Workforce

VOLUME 2

Phil Andrews
Jim Playfoot

ELSEVIER

AMSTERDAM • BOSTON • HEIDELBERG • LONDON • NEW YORK • OXFORD • PARIS
SAN DIEGO • SAN FRANCISCO • SINGAPORE • SYDNEY • TOKYO

Elsevier
Radarweg 29, PO Box 211, 1000 AE Amsterdam, Netherlands
The Boulevard, Langford Lane, Kidlington, Oxford OX5 1GB, UK
225 Wyman Street, Waltham, MA 02451, USA

Notices
Knowledge and best practice in this field are constantly changing. As new research
and experience broaden our understanding, changes in research methods, professional
practices, or medical treatment may become necessary.

Practitioners and researchers must always rely on their own experience and knowledge
in evaluating and using any information, methods, compounds, or experiments
described herein. In using such information or methods they should be mindful of
their own safety and the safety of others, including parties for whom they have a
professional responsibility.

To the fullest extent of the law, neither the Publisher nor the authors, contributors, or
editors, assume any liability for any injury and/or damage to persons or property as a
matter of products liability, negligence or otherwise, or from any use or operation of
any methods, products, instructions, or ideas contained in the material herein.

ISBN: 978-0-12-800975-8

British Library Cataloguing in Publication Data
A catalogue record for this book is available from the British Library

Library of Congress Cataloging-in-Publication Data
A catalog record for this book is available from the Library of Congress

For information on all Elsevier publications
visit our website at http://store.elsevier.com/

  Working together
to grow libraries in
developing countries

www.elsevier.com • www.bookaid.org

Dedication

The authors would like to dedicate this book to all of the highly talented, vocationally trained men and women working in the energy industry globally.

Contents

8.   Wintershall Libyan Integration and Development Programme

9.   The Well Control Institute, USA

About the Authors

PHIL ANDREWS

Most of Phil's career has been as an entrepreneur connected to education and training in the upstream oil and gas business. Having graduated from Nottingham University with a degree in Politics, he was made an Honorary Life Member of the University Union. In 2003, after some travel and a short period working in conferences and publishing, Phil became half of the partnership that founded Getenergy in Aberdeen. He has since been awarded the Livewire Young Entrepreneur of the Year Award in London.

Under Phil's leadership, Getenergy's brand has achieved widespread recognition in the oil and gas and education sectors and more than 40 countries are regularly involved in its meetings, networks and publications. Phil is often asked to speak at international events and is an influential and well-respected international figure in the oil and gas industry who has visited more than 45 countries in support of Getenergy's activities. His opinion is increasingly sought by ministers and their governments in matters connected to developing sustainable economic value through training and education funded by oil and gas activities. In 2014, Phil assumed the role of Chairman of Getenergy Events and is a Non-Executive Director of Getenergy Intelligence.

JIM PLAYFOOT

Jim is a consultant, researcher and writer working in the field of education and skills development. He is the Founder and Managing Director of London-based education consultancy White Loop.

Jim's work over the last 10 years has focussed on understanding the dynamic between education and employment, exploring the challenges of how we prepare young people for the twenty-first century and developing new thinking around how education can have real impact on wellbeing and quality of life. Jim's work is built around a deep understanding of how people learn allied to an ability to engage, analyse and interpret evidence and opinion and produce outputs that are compelling and accessible.

In 2011, he was approached by Getenergy to become involved in developing their research and intelligence function. In 2014, Jim was part of the team that established Getenergy Intelligence and assumed the role of Managing Director, a position that will see him lead on the authoring of all four of the Getenergy Guides volumes.

Preface

Welcome to the Getenergy Guides Series. In this volume, we focus on the challenge of building a technically competent workforce. After finding commercial quantities of hydrocarbons in a 'friendly' country, this could be considered as the most important challenge facing oil and gas companies everywhere. The technical competency of employees lies at the heart of the productivity and safety of every company working in the industry. What is more, the need to develop technically competent workers transcends the oil and gas industry. The degree to which nations remain 'friendly' to the international oil and gas business depends, in part, on how effectively the resource windfall transforms into sustainable economic and employment growth and diversification across the economy. We believe such growth and diversification comes down to the strength of a country's education and training system and its ability to evolve and grow.

However, like any commercially and politically important concept, building a technically competent workforce comes with semantic difficulties, which are best dealt with at the start of this volume, rather than this becoming an unwanted distraction to the reader as the stories we tell unfold. With this in mind, we have decided to explore the concepts to which the title of this volume refers up front.

FIRST WORD – BUILDING

This implies that there is not already something to begin with. Or if there is, it is a set of components awaiting assembly. This is obviously not the case for many of those nations who have been exploiting natural resources for years. However, it is true that nascent energy nations view the industry as a vital lever that can help the transition from an agricultural to an industrial and then, perhaps, a post-industrial 'knowledge' economy. This aspiration is not a remote goal of developing nations but in fact rather mirrors the story of the cities of Aberdeen and Houston, the latter of which features as a case study in this volume. Building the workforce is demonstrably a key component of that development. And as the Houston case demonstrates, rebuilding is often necessary.

In using 'Building' not only are we referencing workforce development underpinned by education and training, but we also want to accentuate the positive opportunity that exists here, (something we originally set out in the Preface of the first Getenergy Guide where we explored the concept of 'Energy, Education and Economy'); namely that whatever the workforce might look like today, whatever its demographics, adaptability and size, the workforce of any nation is continually renewed and rebuilt. There's always more to do and more that can and must be done.

Having chosen the word 'Building' we then necessarily enter the realm of 'architecture' and 'project management'. In the case of a workforce, this inspires a pivotal question: Whose responsibility is it to build a workforce? The State? The education providers? The companies hiring people? The parents of those who will comprise the workforce? Or the individuals themselves? Given that the answer could be any or all of the above, we have decided to pursue case studies in this volume which variously demonstrate not whom should be undertaking the building job, but rather who benefits. In our Caspian Technical Training Centre case, it is clear that BP have benefitted significantly from their investment in building a technical training centre to meet their workforce needs. Estimates indicate a $500 mn USD saving against the cost of employing foreign workers. And that is just the economic cost – we also encourage our readers to weigh up the political and social advantage from investing in the development of a local, competent workforce. The impact of Wintershall Libya's LID Programme – which we outline in this volume – is an excellence case in point.

SECOND WORD - TECHNICALLY

The technical competence question is vitally important to understanding what this book and its cases seek to demonstrate. Namely, that the oil and gas industry is a technical, high-risk, high-cost business. In commissioning and building projects, it is the technical roles which demand many thousands of people, all of whom are critically important to the safe and efficient deployment of infrastructure and the development and operation of assets. Whether offshore or onshore, deep water or desert shale, we hope we will not encounter too many people who take issue with us on the technical focus the cases in this book pursue.

But this does lead to an allied and equally demanding question from our education colleagues. When are skills technical? What is the difference between technical and vocational skills? Surely geoscience graduates are technically skilled as much as a person with a welding ticket? So if the question to answer is 'what do you mean by technically?' our response is simple. In this volume, we are focussing on the greatest component within any workforce in nations with significant industrial expansion; namely people who complete practical tasks with their hands, or, who complete otherwise labour-intensive tasks with computers, which would once have been done by hand. This could, of course, equally apply to writing a book (as an example that is close to hand) but we are confident that our readers will understand what we mean!

To further bring clarity to this topic, we should also highlight that in exploring technical competency, we are largely looking at what happens within colleges of technical and vocational education, what is delivered in-house by international and national oil companies and what is offered by private training providers. We're also keen to point out our focus on non-degree-level education. However, our exploration of the vocational education system in Norway demonstrates how the lines can somctimes become blurred between vocational and academic.

THIRD WORD - COMPETENT

According to the Oxford English Dictionary a person can be competent if they have 'the necessary ability, knowledge or skill to do something successfully'. Without getting into the debate about whether anyone can aspire to such perfection – and what constitutes 'success' – the notion of 'competent' is, to put it mildly, loaded. There are other words that can be applied to elevate still further this concept; perhaps an oil/gas industry definition might read: 'the necessary ability, knowledge or skill to do something successfully, safely and sustainably in environmentally and commercially sensitive conditions'. We could go on. Indeed, a large section of the commercial training and qualifications industry does go on about competence at great length. Why? A variety of explanations offer themselves, but it is quite possible that since the oil and gas industry started widely employing the competency lexicon 25 years ago, training businesses have cottoned on to it as a means of describing the value they add to the companies and their employees. Or perhaps, under pressure in the 1990s to demonstrate they add value, the training industry taught 'competence' to the oil companies. Who knows? But the result has been dramatic. With each disaster, near miss or minor technical problem, the question of proving competence is raised to the executive level. Anecdotally, we understand that a senior oil executive was required to give evidence to the UK Parliament concerning a recent oil spill. Before he could even introduce himself to the committee he was asked by one rather determined Member, whether he could prove himself competent to answer the ensuing questions. He could not. So 'competence' is a word that is used widely, often without its connotations being fully understood.

We must remember that 'competence' is also a source of very significant commercial and financial advantage to those companies and organisations who seek to demonstrate their ability to define and measure it in others. In one of our cases we examine the creation and remit of the Well Control Institute (WCI) in response to the Macondo disaster in the Gulf of Mexico in 2010. The WCI was intended to be the industry's response to this most serious incident and the organisation was tasked with developing and promoting a global standard for drilling and well control training in the upstream business. In its attempt to coalesce the global well control industry around a single – and laudable – objective (to improve competence and safety in well control) we can begin to understand the challenges that exist in collectively defining what 'competence' means in practice.

The concept of 'competence' is inherently linked to another of those education and training buzzwords – 'standards'. We know we are competent when we reach the requisite standard. But this then begs the question – who defines and owns the standard? We spend a lot of time with people who talk wistfully about having a global standard of one kind or another for the oil and gas industry. But in a locally based workforce in developing nations across the globe, how relevant and applicable is any standard conceived in the United States, Europe or the UK? We are aware of one International Oil Company whose response to the Macondo disaster was to assemble a team in their European headquarters and spend two years writing a single global standard for their operations. However, there is a fundamental problem with the notion of ownership when it comes to standards particularly when we rely on those standards to define and measure competence on a global basis. The idea that a single oil company can impose a standard on a workforce when the future of its business relies on employing people who come from a local education system (with its own curriculum, teaching methods and a legitimate claim to work in the best interests of citizens) can be problematic. The approach taken in Kazakhstan (which we consider in our case on Kasipkor) is to qualify candidates locally, award them a national certificate and give them the opportunity to also qualify with international accreditation. Perhaps this is a more balanced approach that recognises the cultural, economic and social aspects of what it means to be competent.

SO CAN WE BUILD A TECHNICALLY COMPETENT WORKFORCE FOR THE FUTURE OF THE GLOBAL OIL AND GAS INDUSTRY? (AND HOW WILL THIS BOOK HELP?)

In the cases in this volume, we have sought to explore intensely practical examples of the efforts which oil and gas companies, contractors, educational institutions and governments have made in recent years to build a technically competent workforce. We regard this as an essential part of the core business of oil and gas companies operating in today's high cost/high risk environment and we hope that this book will help highlight the approaches which work and offer a framework against which to measure future initiatives.

We believe that the challenge of building a technically competent workforce can only be met through collaboration and partnership at a local level coupled with an open approach to the role of international education and training providers. The stories we explore in this book are a testament to that reality and demonstrate one further truth - that sharing our successes and failures (and learning from both) is a positive and constructive way forward.

The process of building a technically competent workforce cannot be successful if done in isolation from the education systems that will be the source of this industry's employees in the next 25 years. To be effective, any competence measure or standard must be inclusive, independently managed and fair. It must be developed as a true partnership with the local colleges, polytechnics and universities in resource-owning countries if it is to assure oil and gas companies that they can employ local people – the work of the community colleges in Texas stands as a pertinent example (and one which we cover in our case on Houston). The role and support that international training and education organisations can provide must be integrated into a model which allows standards to be incorporated into the technical and vocational curriculum. If the industry wants people emerging from a national education system who are 'field-ready' it must be prepared to listen carefully to, and support, the faculty and leadership of institutions in the places where oil and gas are in the ground and from where the future workforce will inevitably come.

We would like to express our gratitude to those organisations, governments and individuals whose time and effort has contributed to this book and whose experiences are reflected throughout the nine case studies we explore.

Phil Andrews and Jim Playfoot
Getenergy Ltd
December 2014

Acknowledgements

We would like to thank the team that has helped us produce this book.

- Getenergy's Cofounder Peter Mackenzie Smith who read and reviewed these pages
- The Getenergy team in London: Helen Jones, Christina Westlake, Jack Pegram, Frankie Carlin-Barrett, Richard Harmon, Annamika Porter-Sinclair, Nick Cressey, Conchi Perez and Tom Fraser
- Simon Augustus for his dedication and hard work
- Special thanks to the talented team at Digital Storm whose cover design captured the theme of this book - www.digitalstorm.com

We would also like to extend profound thanks all the contributors who have given up their time and shared their stories with us.

The Case Studies

The following case studies have been carefully selected to demonstrate the approaches that different organisations have taken to addressing the challenges of building a technically competent workforce. We have tried to present varied narratives from around the world, to reflect their nuances and to tell these stories as honestly and accurately as we can.

Every case study has been developed primarily through in-depth interviews with those involved and rigorous desk research in order to establish the facts. We have, in every case, sought to uncover not only the successes but also the challenges and failures. Where possible, we have included levels of investment and data relating to impact. We have attempted, with each case, to give an entirely accurate reflection of the story and to reflect the contributions of those who took part.

Following each case study, we have included analysis from the Getenergy team – this is designed to give the reader a deeper insight into the case and to uncover what we can learn and take away from each case. It should be made clear that the 'Getenergy View' represents the opinions of our editorial team and not the inputs of our case contributors (to whom we are eternally grateful). Getenergy offers an independent view. It does not itself engage directly in any education and training delivery.

Every case study also includes comment relating to three key criteria: replicability (the degree to which the approach would be replicable elsewhere); sustainability (the degree to which the approach is likely to be sustainable over time); impact (the number of learners/employees educated or trained and the impact on the company or on the industry as a whole). Once again, these observations have been provided by the editorial team at Getenergy and are designed to offer further analysis of each approach.

About Getenergy

Established in 2004, Getenergy is now an independent, privately-held group of organisations that brings together providers of education and training (universities, colleges and private providers) with national and international upstream oil and gas companies, governments and service providers.

Getenergy's mission is to bring new intelligence to the development of skills and competence for the energy industry. This mission is underpinned by the belief that the only long-term solution to creating a sustainable global energy industry is local capability in the exploration, production and transport of energy. Getenergy connects, without barriers, energy companies, governments and educators to achieve this.

In 2014, Getenergy launched a new company – Getenergy Intelligence. Working alongside the existing Getenergy Events business - which designs and operates meetings that create links between education provision and the needs of the oil and gas industry - Getenergy Intelligence aims to capture the stories, ideas, opinions, data and case studies that are at the heart of every engagement we have. The desire to capture, distil and share global good practice is made real through the publication of the Getenergy Guides Series and reflects our mission to support and promote the development of effective education and training in hydrocarbon-producing countries. We work with a select group of knowledge partners in order to guide the focus of our work and to help in disseminating our content to the widest possible audience. We are particularly focused on working with these partners to build local education and training capacity in oil- and gas-producing countries across the world.

Case Study 1

The Norwegian Approach to Competency Development in the Energy Industry

From School to Industry – How the Norwegian Model of Technical and Vocational Education has Nationalised the Energy Workforce

Chapter Outline

With thanks to Hanne Grethe Kvamsø, CEO, OOF and Thina Hagen, Expertise Development Manager, Norsk Olje Gass

THE MOTIVATION

Following the first discovery of natural gas on the Norwegian continental shelf in the late 1950s, it became evident that the North Sea around the coast of Norway had significant potential for oil and gas deposits. At this time, energy needs in Norway were met predominantly by coal and imported oil. Norway now had the opportunity to develop a national industry that could offer energy self-sufficiency and that would have a significant impact on Norway's future economic output.

As drilling began on a commercial scale in the mid-1960s, Norway had no national energy company and no structure of education and training within which to educate and develop nationals to work in the industry. Although the drive to commercialise available resources was strong, successive governments resisted the temptation to sell off exploration rights. In 1963, Einar Gerhardsen's government proclaimed sovereignty over the Norwegian continental shelf and put in place legislation that would ensure that all available natural resources would be carefully controlled at a national level.

Education and Training for the Oil and Gas Industry: Building A Technically Competent Workforce.
http://dx.doi.org/10.1016/B978-0-12-800975-8.00001-0

Initially, those employed in the nascent Norwegian energy industry came predominantly from the UK, Italy and the USA – all countries with established energy sectors and the education and training systems in place to support workforce development for the industries both at home and abroad. In the early years of the industry in Norway, ownership became a key issue and one that extended not only to the physical boundaries of the continental shelf but also to the culture and education of the workforce.

With the establishment of Statoil and the recognition of the size and scale of the industry (in what is a relatively small country in terms of population) the potential impact on employment and on the wider economy was clear. There was a need to create a coherent approach to workforce development that would underpin the gradual nationalisation of the workforce and that would ensure Norwegians assumed control of an industry that was to shape the economic future of the country.

THE CONTEXT

The Energy Industry in Norway

In the late 1950s, it was not thought that the Norwegian continental shelf concealed any significant quantities of oil and gas deposits. The discovery of natural gas at Groningen in the Netherlands in 1959 caused people to revise their thinking on the potential that existed in Norwegian waters for hydrocarbon production. This discovery generated enthusiasm for energy exploration in a part of the world where much of the energy consumed was imported and where there was no national energy industry to speak of. Although there was an initial caution towards the opportunities that the North Sea might offer for hydrocarbon extraction – with geologists in Norway sceptical of the potential – others believed that, after the gas discovery in the Netherlands, the Norwegian continental shelf may hold significant reserves.

In October 1962, Phillips Petroleum applied to the Norwegian authorities for an exploration licence in the North Sea. The company wanted to explore parts of the North Sea that were in Norwegian territory and that included the Norwegian coastal shelf. The company attempted to secure exclusive rights for their exploration activities, but this was resisted by the government of the day, in recognition of the downside of awarding responsibility for the whole shelf to one company.

In May 1963, Einar Gerhardsen's government proclaimed sovereignty over the Norwegian continental shelf and began to implement regulations that established the State as owner of all natural resources within Norwegian territory. This meant that the government was now responsible for awarding all licences for exploration and production. The same year, a number of companies were awarded the rights to carry out preparatory exploration, although these licences only extended to seismic surveys, not to drilling.

The Ekofisk discovery in 1969 represented the first major commercial find in Norwegian territory and was the first discovery of oil to result from the extensive exploratory drilling activity in the North Sea that ensued after the Groningen natural gas find a decade earlier.[1] Production from the field started in 1971 and was followed by a number of major discoveries within Norwegian waters.

At this time, Norway remained outside the European Union and also took the decision to not join OPEC and, as a consequence, to set its own energy prices. As the industry grew, international companies were dominating exploration and were responsible for developing the country's first oil and gas fields. This reflected the nascent nature of the Norwegian oil and gas industry, where no national companies of any size or stature existed. Within this context, Statoil was set up in 1972 as the national energy company of Norway. At the same time, the government established the principal of 50% local participation in every production licence. This was, in effect, the beginning of Norway's local content policy for the industry. The government also chose to award drilling and production rights to Norsk Hydro and Saga Petroleum, both companies wholly owned by Norwegians and illustrative of the gradual growth of the national energy industry in Norway at that time.

An oil rig set in a Norwegian fjord.

In 1985, government participation in petroleum operations was reorganised with the State continuing to own interests in a number of oil and gas fields, pipelines and onshore facilities alongside the private ownership by

---

1. Van den Bark, E., and Thomas, O.D., 1980, Ekofisk: First of the Giant Oil Fields in Western Europe, in Giant Oil and Gas Fields of the Decade: 1968–1978, AAPG Memoir 30, Tulsa, American Association of Petroleum Geologists, pp. 195–197.

international oil and gas companies. The size of the government stake was decided when production licences were awarded and the size varied from field to field. As one of several owners, the State would pay its share of investments and costs and would then receive a corresponding share of the income from the production licence. In 2001, Statoil was listed on the stock market and effectively privatised. It now operates on the same terms as every other commercial company on the Norwegian coastal shelf. Petoro was established in May 2001 as a state-owned limited company to manage the Norwegian government's oil and gas interests.

Petroleum activities have had a profound effect on economic growth in Norway and on the financing of the Norwegian welfare state. Over the course of 40 years of operations, the industry has created revenues in excess of NOK 9000 billion.[2] In 2012, the petroleum sector accounted for 23% of value creation in the country.[3] Since the petroleum industry started exploration activities in Norwegian territory, significant investments have been made in exploration, field development, transport infrastructure and land facilities. At the end of 2012, this investment was estimated to be around $480 billion USD in current terms. Investments in 2012 alone amounted to over $28 billion, or 29% of the country's total investments. The industry accounts for around 1.5% of all employment.

An oil platform in Tromso, Norway.

2. Norwegian Petroleum Directorate – http://www.npd.no/en/Publications/Facts/Facts-2012/Chapter-3/.
3. EIA – http://www.eia.gov/countries/cab.cfm?fips=no.

In spite of more than 40 years of production, it is thought that only 42% of the total reserves on the Norwegian continental shelf have been exploited.[4] Production reached a peak in 2001 of 3.4 million barrels per day and gradually declined over the following decade.[5] In 2012, petroleum production was 1.8 million barrels per day.[6] The volume of gas sold the same year was 114.8 billion cubic meters.[7] There are 8000 km of offshore gas pipelines running across the Norwegian continental shelf with landing points in four countries in Europe. Currently there are 53 licensee companies on the Norwegian continental shelf and 42 exploration wells were drilled in 2012, demonstrating that the industry continues to thrive and grow.[8]

In 2011, there were 76,848 people employed in the petroleum and petroleum-related industries. Of these, 72,263 were settled in Norway, while 4585 were settled in other countries and were working in Norway on a temporary basis. Between 2011 and 2012, employment across the industry increased by around 7500 – an increase of around 10%. Employed foreign residents increased by around 14% between 2011 and 2012, slightly down from 15% from 2010 to 2011.[9] These figures serve to demonstrate that the demand for skilled workers remains strong and that imported and temporary labour still accounts for a significant proportion of the workforce.

The Need for an Education-Driven Response

As Norway began to recognise the impact that hydrocarbon discovery in the North Sea was going to have on the economy, there was also a realisation that the emergent energy industry could be a driver for employment, entrepreneurship and national business growth. At this time, it was also evident that education and training would become critical in terms of capitalising on this new-found mineral wealth. The drivers for creating a new education and training system that could meet the demands of industry were clear:

- Without educating and training local people, jobs at every level across the industry would remain the domain of expatriate workers, meaning that there would be no significant employment dividend from what would become a critically important national industry.
- By educating and training local people, the industry would gradually become nationalised, thereby promoting Norwegian companies as key players and

---

4. Petroleum Resources on the Norwegian Continental Shelf, 2014.
5. EIA – http://www.eia.gov/countries/cab.cfm?fips=no.
6. Ibid.
7. Bloomberg, May 2, 2013. *Statoil profit slides more than estimated as production declines.*
8. Norwegian Petroleum Directorate – http://www.npd.no/en/news/News/2013/The-Shelf-in-2012-press-releases/Exploration/.
9. All figures from Statistic Norway.

driving forward the industry in a way that would be beneficial to the country as a whole and not just to shareholders of international companies.

- By educating and developing suitably trained nationals, the workforce would become permeated by Norwegian culture, furthering the interests of the country and its people and enabling the industry to develop in a way that reflected Norwegian values.
- The newly formed national energy company – Statoil – needed Norwegians who were capable, well-trained and able to assume senior positions in the company based on their knowledge and experience of the industry.
- Exploration and production in the North Sea presented many technological challenges and Norwegian companies needed to invest in building the technical capabilities to meet these challenges or be left behind by international competitors.
- There were parts of Norway that were in need of industrial and employment regeneration following the demise of the shipbuilding industry (particularly in Stavanger and Oslo). The energy industry offered the opportunity for this regeneration assuming that the local population were able to gain the requisite competencies to do the job.

In 2014, there are around 24,000 people working on the Norwegian continental shelf.[10] Despite the success of the education and training system that has been in place for two decades (and that we explore further below), the need for a demand-driven approach to workforce development remains. The industry in Norway is currently facing three key challenges:

- the increasing need for personnel generated by ongoing levels of exploration and production activity,
- growing international competition over the best qualified and highest performing employees and
- a demographic challenge that is seeing baby boomers gradually retire and needing to be replaced by new recruits.

These challenges continue to define the development of Norway's vocational education and training (VET) system and the impact this has on the energy sector.

THE SOLUTION

During the 1970s – when the energy industry was expanding and oil and gas companies were beginning to employ Norwegians – Norway began to implement dedicated higher-education programmes in petroleum engineering. Today, these programmes are well established and highly respected around the world. However, the bulk of employment within the industry resides in technician-level jobs and these require vocational training.

---

10. Aljazeera, 18 December, 2006. *Norwegian oil companies join forces.*

The Petroleum Museum in Stavanger.

As the industry grew, those recruited into offshore positions were typically coming into the industry directly out of school with no formal qualifications and without the skills and competencies required to do the job for which they were being employed. These candidates would then undertake equipment-based training that would be provided by the company that employed them. Most drilling and well-services personnel undertook 420 hour drilling courses delivered by providers who largely certified themselves. This meant that, as far as skills development and workforce planning was concerned, the state education and training system had little or no role to play in supporting the industry or in developing individuals to fulfil positions in the industry. The training and development provided was largely ad hoc. Such a system failed to adequately meet the needs of industry and did not afford local candidates the opportunities they deserved for career or educational progression.

The solution was a demand-led, industry-focused technical and vocational system that prepared school-leavers for working in the industry and that linked their school-based education to the job that they planned to do. In the early 1990s, Norway instigated a programme of education reform that supported young people in upper-secondary-level education to develop a vocation. Since then, young Norwegians have had the right to undertake an upper secondary education whereby they can choose between three general studies courses and nine vocational courses. Most young people attend upper secondary school at the age of 16 years and more than 50% of those enrolled apply to VET programmes (with the remaining 50% undertaking one of the three general studies programmes). Drilling and well-services courses are now part of the vocational stream and this has been an important step in the development of a nationalised workforce for the industry.

The approach adopted is underpinned by the belief that a successful economy is based on an education system that is fit for purpose and that gives opportunity to everyone. The system also actively promotes vocational study as a worthy and worthwhile pathway – something that many other countries fail to do.

The challenge for the school system in Norway – particularly in relation to energy-related courses – has been to balance the school-based learning with what employers in the industry are best placed to provide. The Norwegian system recognises the fact that there needs to be a balance between industry requirements for cutting-edge expertise and the needs of the society to have a well-educated population with good levels of general competence. Within this context, specialised industry training (for which there will always be a need) is considered a matter for individual companies to address; the state education system seeks to provide a basic level of technical expertise that will serve learners well in their future career and develop a pool of talent that the country can benefit from.

The Norwegian model for technical and vocational education – as applied to the oil and gas sector as well as other vocations – is characterised by the following aspects:

- The system is a close collaboration between all key partners – this includes employers, unions and educational authorities at both national and local level.
- The system is driven by this collaboration and it underpins all critical aspects of what is done, including defining the need, developing the qualifications, designing the assessment and delivering the learning experience.
- 16-year-olds who are interested in working in the industry can choose a vocational course that requires them to follow a dual system of education and work-based learning via an apprenticeship.
- For those looking to work in well-control, first-year students attend one of two VET programmes – a programme for technical and industrial production or a programme for electricity and electronics. The following year, they specialise and attend in-depth courses on specific aspects of their chosen specialism. Health, safety and environmental training is a part of the teaching across all courses.
- After 2 years at school (the first 2 years of the course), students are required to pick a specialisation that they will carry into their third (apprenticeship) year.
- All courses involve undertaking apprenticeship training with companies, some of which will be offshore. Companies that take apprentices on have to be approved as a training enterprise by the education authorities before undertaking that role.
- All apprentices are subject to continuous assessment in work and this is a critical part of their professional development. This requires the employer to be on board and willing to undertake the assessment in a professional and proactive manner.
- Industry involvement in the system is the key: the industry has a significant role to play in influencing the content of what is taught in schools and then assumes a key role as the apprenticeship provider. The syllabuses that are developed by employers and other industry bodies are enshrined in statutory regulations and schools and other training establishments are bound by their content.
- All of those graduating from this process receive a recognised trade certificate. Final assessments are undertaken by company representatives

to ensure continuity – this further demonstrates the way in which industry is critical to every stage of the process (including what trainees are examined on).

- Training progression is a key part of the system – those who receive their professional certificates are able to go out and work in the industry at an appropriate level or they can choose to further develop their skills by pursuing other qualifications. This pathway leads right from the professional certificate received at school up to Ph.D. level. Those choosing to work immediately on completion of their qualification are able to return to education at any stage.
- For those who are already working in the industry but want their skills and competencies to be recognised, there is an alternative pathway they can follow. The 'experience-based trade certificate' is designed for those with more than 5 years of industry experience. Existing workers follow the same apprenticeship curriculum as new recruits and complete the same exams. This enables these individuals to have prior learning and competency recognised and offers them the same progression routes as non-employed trainees/apprentices.
- In Norwegian higher education, all vocationally oriented courses and programmes are part of the ordinary higher education system. There is no formal or other distinction between vocational and non-vocational higher education. This is a contributing factor to the elevated status of technical and vocational study in Norway.
- As a publicly run system, the central bodies that run the system – which include the Norwegian Oil and Gas Association and the National Training Office for Oil Related Trades – have a very strong influence over the way in which the programmes are delivered. Classroom teachers receive training in the curriculum that the apprentices will experience as they undertake their apprenticeship and there is significant focus on developing the right behaviours for work (which is a critical part of any job but particularly important within the energy sector).

There are seven drilling and well subjects available to students in their third year of study: cementation, coiled tubing, completions, electrical cable operations, mechanical cable operations, seafloor installations and drilling/roughnecking. There is also the opportunity to undertake a qualification in remote-operated vehicle operations.

All qualifications are developed in a tripartite cooperation between the industry, the unions and employer organisations and the school authorities and all qualifications are updated on a regular basis via industry advisory panels.

Although the work-based learning component is vital, much of the learning experience happens in the classroom. This means that the quality of teaching is vital to the overall success of the programmes delivered. The drilling and well subjects are taught by qualified teachers. However, many have a background in the oil and gas industry (typically as geologists or engineers). Those working in the industry are able to undertake teaching courses, paid for by the government, if they want to work as teachers.

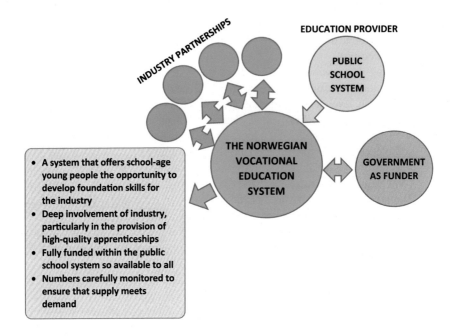

The system is underpinned by a proactive approach to workforce planning that involves two key elements:

- At the age of 15, all young people go to vocational fairs where they can explore and discover a range of different careers that are available to them (including careers in the energy industry). Their parents are also encouraged to be involved at this stage, as parental concern about the experience of working in the energy industry can be a barrier to attracting talent and these events give industry representatives an opportunity to address those fears.
- The government invests continuously in collating and updating labour market information to ensure that there is a clear picture of industry demand and that the supply side is meeting (and not surpassing) this demand. As a consequence, the state is able to manage the numbers going into specific trades very carefully so that industry needs are met and there are not too many people completing each course. This also means that places on specific courses are limited and very competitive.

The Norwegians have managed to develop a model that produces high-quality graduates, in partnership with industry, and delivered within the public school system. This means that the experience is accessible to all and as the system is not run on a commercial basis, the focus of those involved is entirely on achieving the best outcomes for learners and for the industry.

Education and training within the energy sector is significantly influenced by the work of the Education Office of Oil Related Sciences (OOF). This organisation was established in 1999 and its remit is to ensure that the oil and gas industry supports the development of future professional expertise by working with education authorities to implement an effective vocational training system for the industry. This organisation performs a number of key roles and acts as a bridge between the education system and the industry. Their activities include the following:

- Assisting companies looking to take on apprentices and organising the application process.
- Ensuring that companies meet their responsibilities in regard of apprenticeships.
- Arranging contracts and other documentation between education institutions and the industry.
- Following up apprentice training within the business and ensuring that apprentices receive training that meets the requirements of the curriculum.
- Ensuring that in-work assessments are carried out effectively.
- Ensuring that trainees have access to the right courses that develop the skills required to pass the qualifying examination.
- Promoting the teaching of subjects relevant to the industry in schools.
- Informing the vocational training board of any changes in learning conditions within inspected schools.
- Mentoring companies and apprentices in the rights and obligations they have, and ensuring that agreements, laws and regulations are fully understood.
- Keeping up to date with legislative changes and with any information produced by ministries, local authorities and other organisations that have relevance to subjects taught.

Of particular importance here is the work that OOF does in ensuring that the apprenticeship scheme that underpins the education and training programmes is effectively run and managed. The challenges associated with running a successful apprenticeship system should not be underestimated and the focus given to this task within OOF's remit is testament to this.

OOF operates as a training provider delivering a range of industry-relevant courses. OOF also offers learners career advice and helps learners to explore their options once they have completed their studies. By acting as a bridge between education and industry, OOF plays a significant role in ensuring that candidates get the right advice, the support they need and access to vital work-based learning opportunities. They are a critical part of the education and skills ecosystem within the energy industry in Norway.

THE IMPACT

This system has now been in place for some years and has had a profound effect on the talent pool coming into the industry. The benefits can be seen in the following ways:

- This system is geared around industry involvement, so those emerging from it have the core competences needed by industry to fulfil workforce requirements.

- The model enables companies working across the industry to forge links with talented young people from an early stage. Through the offering and completion of apprenticeships, many students go on to work for the company where they studied, thereby ensuring a smooth pathway into employment for the candidate and a much reduced recruitment cost for the company.
- Candidates are given a broad and rounded education that develops a good level of basic professional competence. They are then able to specialise in a particular field, but their basic competency is something that will serve them well regardless of the particular area of business they end up in. This means they have a greater set of professional options available to them when they leave school.
- The significant focus on work-based apprenticeships ensures that candidates are well prepared for the world and have a good understanding of what to expect when they leave school and assume full-time employment.
- There are currently around 400 apprentices taken on within the industry every year. These numbers are highly controlled in order to ensure the right number of graduates for each area of the business.
- Through many years of investment and an open approach to collaboration, the Norwegian system has emerged as a model of how public authorities can work together with industry and with education and training providers to develop skilled and competent future employees. What is particularly impressive is that this takes place within (and across) the public school system.

The most recent figures available suggest that around 95% of those employed in the industry in Norway are settled residents.[11] Although a significant number are non-Norwegians, this suggests a degree of success in terms of developing a stable, committed local workforce for the industry. Anecdotally, conversion rates from apprenticeship to employment are extremely high for those candidates who wish to join the labour market on completing their initial qualification. For others, pathways to higher and further education are readily available. A recent study also suggested that a larger proportion of employees in the petroleum and petroleum-related industries had a higher education compared to the rest of Norway's private sector. This is also indicative of the success of this approach.[12]

THE CHALLENGES

A number of key challenges have been identified by those involved in shaping the education system in Norway:

- Of the 400 apprentices taken into companies every year, only around 5% are women.[13] Efforts are being made to increase this number, but the challenge is

---

11. Figures from Statistic Norway.
12. Employment in the petroleum industry and related industries, Statistics Norway, 2014.
13. Figure supplied by OOF.

significant in attracting women into the industry and demonstrating to them that it can offer a fulfilling career.

- Industry skill requirements for offshore workers currently exceed what the education system is producing. Although there is an expectation that the industry will offer new hires additional training, there is a need to constantly monitor the quality and relevance of what is taught in the vocational system and improve where necessary.
- The demands placed on industry partners is high: the required ratio of apprentice to assessor is one to one and when trainees go offshore, they are required to have an evaluation with their instructor after each trip. This is a resource-heavy approach to training and development and one that requires significant buy-in from the industry.
- It remains a challenge to find the right type of employers able to offer enough good-quality apprenticeships every year. The quality of the apprenticeship experience is such an integral part of the process that this has to be a priority for the various authorities that run and manage the system.
- It could be said that the approach has become a victim of its own success. With international competition for skilled, technically trained workers becoming ever fiercer, the effectiveness of the Norwegian system means that graduates are in high demand and the industry in Norway now has to fight hard to retain their own people.

THE COST

The apprenticeship scheme – which arguably lies at the heart of the success of the system – is directly funded by government subsidy with each company receiving just over 100,000 kroner (around £10,000 pounds) a year for accepting and training one apprentice, with many companies taking on more than one apprentice. Although this subsidy is generous, it is unlikely to cover all the associated costs of supporting an apprentice for a year, so in-kind investment on the part of the company is a critical element in the funding model.

The system is funded predominantly by the government (as it operates within the public schooling system). This investment is made possible in part by the ongoing revenues from the energy industry. This makes the Norwegian system a prime example of how a resource-rich country can reinvest the energy dividend in a way that not only feeds the industry with talent but also creates a broader set of education and training benefits for the country as a whole.

Government funding for the system is critical – without major investment over a sustained period the system would not work. In 2008, Norway spent 5% of its GDP on primary and secondary education and training as a whole – the average in OECD countries is 3.8%.[14] Although Norway retains significant reserves of hydrocarbons (and has the finances to continue funding at this level

---

14. OECD, *Education at a Glance 2008 OECD Indicators: OECD Indicators*.

for the foreseeable future), it may ultimately become difficult to maintain a system at this level of funding.

The Getenergy View

- The careful management of private sector participation in education has been key to the success of the approach. Employers are central to the model of apprenticeship-driven education and training that the system follows. Every employer who wants to take on an apprentice must become an accredited training provider in order to demonstrate they have what it takes to develop an apprentice.
- The approach is based around a clear understanding of how far the student can be taken within the school system before a company needs to assume an active role in their continued educational development. The approach sets out to support the development of basic skills and then assumes that specialised training will be provided by industry. This suits both the public school system and the recipient companies.
- Competition is essential. The oil and gas industry will not take just anyone – they want to take the best. The fact that numbers are tightly monitored in terms of those undertaking courses within the industry means that competition for places is high. This ensures that there is a selection process for candidates that pushes the standard up across the system.
- The approach reflects (and relies on) a high level of commitment from the energy industry, as evidenced by the fact that companies invest in taking apprentices offshore on to rigs where they are exposed to a high-risk environment. This is costly and demonstrates the lengths that companies are required to go in order to meet with the requirements of the training programmes.
- The demand and supply of people is hard to assess and forecast. Norway has a significant amount of data available, but a company's personnel needs change over time and this must be monitored. The success of the system shows how the more data we have, the closer we get to meeting the need.
- The system has implemented mechanisms to ensure that those already working in the industry (perhaps those who received informal training and have learnt on the job) are also able to certify themselves to the same standards as new recruits. Recognition of prior learning is essential as a mechanism to deal with a workforce that historically has not been formally trained.
- Norway is now educating and developing people very successfully, but this may not be replicable in other countries. It may be challenging to motivate companies to get involved in secondary education.
- A critical aspect of the success of the system is that there is no value distinction between vocational and academic – the progression routes remain the same and students are able to seamlessly transfer between both streams.

A Note on Sustainability

The sustainability of the education system in Norway relies heavily on government funding as the provision is entirely in the public school system. As such, we could question the long term viability of such a model. However, the culture of high taxes in Norway coupled with continued oil revenues would suggest that such a system will be operational for many years to come.

A Note on Replicability

Norway is a wealthy, developed, high-taxation country with a comparatively small population. Would this approach – which relies largely on government funding – work in other countries with a different economic context? Furthermore, the reliance on companies to provide significant impetus to the apprenticeship scheme would be difficult to replicate in countries where the link between education and industry was not as well established.

A Note on Impact

Our view is that Norway offers a powerful model for how a resource-rich country can use revenues to create significant impact across the education and training system as well as providing a superb, technically competent workforce for the oil and gas industry. The success of the industry reflects the quality of the workforce.

Case Study 2

Building Human Capacity in Saudi Arabia

The Impact of Government Initiatives on the Oil and Gas Workforce

Chapter Outline

With thanks to Frank Edwards, Consultant and Member of the Getenergy Global Advisory Board

THE MOTIVATION

Saudi Arabia is a country rich in natural resources and its principal economic strength is derived from hydrocarbons; roughly 75% of government revenues and 90% of export earnings come from the energy industry.[1] The country was the world's biggest producer and exporter of oil in 2012 and, as of 2014, has oil reserves that equate to about one-fifth of the world's total proven reserves.[2] The wider industrial sector is also expanding, creating additional employment opportunities but also highlighting the need to train a new generation of skilled nationals in order to meet workforce requirements.

The drive towards 'Saudisation' – the process of educating and training Saudi nationals to assume positions within the workforce – has roots in two key factors. First, the country has a young and increasingly well-educated population who have ambitions to work and earn a living wage. Second, unemployment

---

1. Central Intelligence Agency, Office of Public Affairs, *The World Factbook 2008.*
2. EIA – http://www.eia.gov/todayinenergy/detail.cfm?id=10231.

Education and Training for the Oil and Gas Industry: Building A Technically Competent Workforce.
http://dx.doi.org/10.1016/B978-0-12-800975-8.00002-2

in Saudi Arabia has risen over recent years and stood at 11.5% at the end of 2013 with the figure significantly higher among young people, graduates and women.[3] Reports from the OECD and ILO suggest that the unemployment rate among youth aged 20–24 years stood at just under 40% in 2009.[4] At the same time, the working population of Saudi Arabia is dominated by non-nationals with figures suggesting that around eight million foreigners live and work in Saudi Arabia[5] (with the national workforce estimated to be around half this figure). The political, social and economic imperative to rebalance the workforce in favour of nationals is clear.

Faced with a workforce dominated by non-nationals and a pool of untapped talent among local graduates and school leavers, the Saudisation process began. A number of policies have been enacted by the Saudi Arabian government to promote Saudisation with the target Saudisation rate at 75% for the private sector.[6] Events of the Arab uprisings since 2011 have further highlighted to the Saudi Arabian government the need to address youth and graduate unemployment and to ensure that ambitious Saudis have access to the type of training and employment opportunities that they expect and aspire to. Within this context, the oil and gas industry needed to respond in order to keep pace with the ambitions of the government. As a consequence, the industry has been the recipient of a number of flagship initiatives – driven primarily by the Saudi Arabian Ministry of Labour – that have sought to achieve Saudisation targets across the workforce. These are explored in detail within this case study.

THE CONTEXT

Saudi Arabia has 16% of the world's proven oil reserves, it is the largest exporter of total petroleum liquids in the world and maintains the world's largest crude oil production capacity.[7] More than half of Saudi Arabia's oil reserves are contained in eight fields. The giant Ghawar field, the world's largest oil field with estimated remaining reserves of 75 billion barrels,[8] has more proven oil reserves than all but seven other countries. The country also boasts the world's fifth-largest natural gas reserves,[9] although natural gas production remains limited. The industry as a whole provides significant employment. As an example, the national oil company Saudi Aramco employs over 54,000 people.[10]

---

3. Gulf News, May 2014.
4. Ellen Knickmeyer, 2011. *Idle Kingdom: Saudi Arabia's Youth Unemployment Woes Go Far Deeper Than Most Realize*. Foreign Policy.
5. 'New plan to nab illegals revealed'. Arab News. 16 April 2013.
6. According to Article 45 of the Labour and Workman Law of Saudi Arabia.
7. EIA – http://www.eia.gov/countries/country-data.cfm?fips=sa.
8. EIA – http://www.eia.gov/countries/country-data.cfm?fips=sa.
9. Ibid.
10. Saudi Aramco Facts & Figures 2013.

Fuel tanks at the Ras Tanura oil terminal, Saudi Arabia.

The Saudi Arabian labour market has historically suffered from a variety of structural imbalances. Most prominent among these is a high dependence on foreign labour coupled with high unemployment levels among Saudi Arabian nationals, especially graduates and young people. Saudi Arabia's employment problems are related to two main factors. First, the country has a long history of employing foreign workers in the private sector, partly as employment conditions are often seen as unattractive to nationals but also as a result of the skills mismatch between local people and requirements in the private sector. Second, Saudi Arabian nationals have traditionally had a strong preference for seeking employment in the public sector due to the favourable pay and working conditions on offer (a pattern mirrored in a number of other Arab states). As the population expands and the public sector contracts, high rates of youth unemployment have created concerns over the social and political impact that may be seen if the situation is not addressed in a systematic and structured way. The solution must involve significantly greater participation by Saudi Arabian nationals in the private sector. Furthermore, there is a recognised need to diversify an economy that is highly dependent on oil and gas and to grow the retail, tourism, hospitality and other service sectors.

The challenges within the labour market – and the evident issues around the employability of Saudi Arabian nationals – have been recognised as having their roots in the historic failures of the technical and vocational education system to adequately prepare candidates for the workplace. What is more, the choice of a vocational education has not been commonly seen by Saudi Arabian students and their families as one that is desirable or positive (reflecting attitudes towards nonacademic education that pervade across the region).

Furthermore, the Saudi Economic Offset Programme – which was first implemented in the 1980s – demonstrated the systemic challenges facing the Saudi Arabian economy in regard to the labour market. This policy was designed to ensure that foreign companies investing in Saudi Arabia and delivering contracts within various economic sectors across the Kingdom were contributing funds to the wider development of the country as a part of their contracts. A significant target for this programme was to increase the impact and effectiveness of education and training for nationals and, as a consequence, to grow the number of Saudis directly employed by these foreign contractors. However, analysis of this programme suggested that although funds were available for education and training, very little impact was achieved in regard of skills development among the local population and employment rates among Saudis failed to reach target levels. Funding here was not the challenge – the failure of the Saudi Offset Programme in this regard had more to do with a lack of appropriate infrastructure for the training and development of nationals. This experience had a significant impact on the future planning decisions regarding workforce development and can be seen in the initiatives and programmes being implemented now.

The Process of Saudisation

The drive towards Saudisation has been ongoing since early in the twenty-first century (although the Saudi Offset Programme included ideas of Saudisation within its approach). In 2005, the target rate for the number of Saudi Arabian nationals working within the private sector was 75%, although even today, evidence suggests that these numbers remain out of reach within a majority of sectors. In 2011, as part of the ongoing Saudisation programme pursued by the Saudi Arabian Ministry of Labour, the Nitaqat system was introduced. Under the system, private firms are classified into Premium, Green, Yellow and Red categories based on their percentage of Saudisation. The system sets penalties and incentives for companies based on which category they fall under. Companies with high Saudisation rates fall under the Premium or Green categories, while those who fail to achieve the required rates will be included in the Yellow and Red categories. The Nitaqat scheme was designed to ensure that Saudisation rates are based on the actual performance of private businesses and takes into consideration the sector as well as the size of workforce for each company.

The Saudi Arabian Ministry of Labour pledged to take harsh action against general service offices allegedly involved in helping some companies circumvent Saudisation targets. Aside from closely monitoring the market, the Ministry appointed around 1000 inspectors to ensure effective implementation of the Nitaqat programme. There were, however, some indications that companies were fulfilling their quotas by simply employing locals in low-level manual positions (as security guards or drivers) and that this was not contributing to the social mobility of the local population.

Historically Saudisation projects have met with limited success because of entrenched perceptions preventing nationals from pursuing certain types of employment, particularly employment in many parts of the private sector. More pertinently, there have been evident challenges around the mismatch between required skills and the competencies that candidates develop within the education system. For Saudisation to have real impact, this was a challenge that needed to be addressed. As the country set about improving the impact of the Saudisation process, it was clear that the Nitaqat programme (which was essentially a way of policing the compliance of companies in meeting nationalisation targets) would have to be supported by a range of education and training interventions that would address skills development and promote the participation of all citizens. For the oil and gas industry, this has seen the creation of the Saudi Petroleum Services Polytechnic (SPSP), the National Industrial Training Institute (NITI) and the Colleges of Excellence (which are not exclusively dedicated to the energy industry but do educate and train many candidates for the wider supply chain as part of their remit).

Faced with a need to stimulate the private sector, to support the nationalisation of the workforce and to improve the quality and relevance of technical and vocational education, the Saudi Arabian government needed to embark on a significant programme of reform. This programme had to address the systemic issues facing the education system as well as finding mechanisms to connect technical and vocational education more directly to the needs and requirements of industry. As the primary source of revenue and exports, the oil and gas industry would have to be a central element of this programme. Furthermore, the demands of Saudi Aramco – the national oil company of Saudi Arabia – are alone enough to warrant large scale investment in local education and training. Within a context of significant reform in every part of the Saudi Arabian education system, the need was for Technical and Vocational Education and Training (TVET) solutions that could meet the demands of the oil and gas sector now and in the future. This would require a much more proactive approach to partnership, the deep involvement of the industry and a major investment in facilities, infrastructure and teaching that would bring Saudi Arabian technical and vocational provision up to the standards now demanded by national and international oil companies.

THE SOLUTION

The solution to improving Saudi Arabia's TVET system – and to providing a mechanism by which local Saudis could be trained to international standards for jobs in the oil and gas sector – will require a massive expansion of provision. Estimates suggest that as of 2012, only around 10% of Saudis receive any form of technical or vocational training.[11] The ambitions of the Saudi Arabian government are to achieve rates of 40–45%, a fourfold increase.[12] This will mean

---

11. City&Guilds, 2014. *Kingdom of Saudi Arabia invests in skills to boost vocational education provision.*
12. Ibid.

expanding the TVET system from about 110,000 students per year today to more than 450,000 students by 2020.

The solution will also require a recognition that the wholly state-run education system at the higher and further levels has failed to meet the requirements of industry and has been unable to offer the flexible, dynamic, modern TVET programmes that are required in order to develop technically competent workers. To this end, the solutions now being implemented are hugely reliant on public–private partnerships involving international education and training providers.

These solutions are being driven forward primarily by significant government investment into the system through the Technical and Vocational Training Corporation (TVTC) and by strengthening and broadening the powers of the Saudi TVET regulator, the National Centre for Evaluation and Accreditation (which is part of TVTC). The approach adopted by TVTC (which is elaborated below) aims to embrace forward-thinking models of education that take the best of what the international market has to offer and integrates this within an overarching TVET strategy. Within this, there is a clear focus on ensuring international recognition of qualifications, particularly within the petroleum sector. The strategy also recognises the vital need to involve industry deeply in development and provision of qualifications and training. Below, we consider how the government of Saudi Arabia has begun to enact their vision for a more effective, sustainable and impactful TVET system that supports every aspect of industrial activity across the Kingdom.

The Establishment of the Technical and Vocational Training Corporation (TVTC)

A key part of achieving the ambitions set out above has been the establishment of the TVTC. TVTC is the government agency that is largely responsible for the expansion of TVET and is both a training provider and a coordinating organisation for TVET provision. TVTC has undertaken a number of projects designed to improve training capabilities, including establishing new centres and issuing contracts to a number of international training providers. According to TVTC, the organisation has been established to achieve the following objectives[13]:

- Absorb maximum number of students willing to benefit from TVET in order to achieve sustainable development
- Train and develop national manpower in technical and vocational fields according to labour market requirements – both in qualitative and in quantitative terms
- Establish strategic partnerships with industry in order to carry out technical and vocational programmes

---

13. Details provided by TVTC.

- Achieve an even geographical coverage and increase the number and capacity of colleges and institutes (for boys and girls) in all governorates and cities in the Kingdom
- Disseminate awareness among communities about the importance of working in technical and vocational fields, as well as create an appropriate environment for lifelong learning
- Ensure quality when designing and offering training programmes with the aim of gaining national and international accreditation
- Consolidate the relationship with, as well as the integration of, all national educational and training entities

Initially, TVET in Saudi Arabia was run by three different government bodies – the Ministry of Education, the Ministry of Labour and Social Affairs and the Ministry of Municipalities and Rural Affairs. By not recognising the need for clear consolidation of policy in this area, progress on TVET reform was slow and lacked coordination. The establishment of TVTC to serve as an umbrella organisation for all branches of TVET represented the first of a series of steps on behalf of the Saudi Arabian government to address the challenge of building an effective TVET system with a single entity at its heart. TVTC is now responsible for driving policy forward and is also directly responsible for the running of all colleges of technology, girls' higher technical institutes and vocational institutes. In addition to the hands-on role that TVTC plays in the management of technical and vocational institutions, the organisation also has responsibilities for the following activities:

- Designing and implementing TVET programmes
- Conducting TVET research
- Developing and reviewing TVET plans and strategies in accordance with national policies and frameworks
- Qualifying TVET teachers and trainers
- Setting standards, issuing licences and supervising the establishment of private TVET institutions
- Establishing strategic partnerships with training organisations to run and manage TVET institutions
- Advising the public and private sector on TVET
- Designing and developing TVET training tools and technologies
- Participating in national and international TVET cooperation
- Developing the best practices of rules and regulations in TVET

TVTC has recognised that there must be a relationship between higher education and technical and vocational training – as such there is an emphasis on linking both education levels, as detailed by the National Qualifications Framework for Higher Education in Saudi Arabia (2009). Thus, the granting of credits and allowing exemptions from programme requirements for students wanting to change their education path have been implemented.

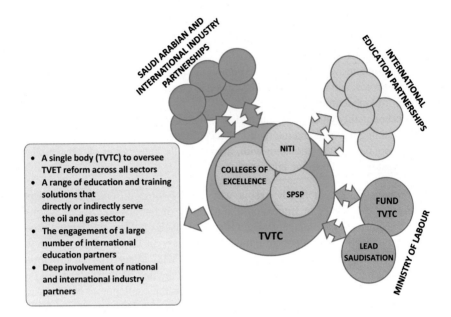

The development of National Occupational Skill Standards – a process led by Saudi Skills Standards, an offshoot of TVTC, and the National Centre for Evaluation and Accreditation – has helped improve occupational certification, training and employment and has enabled the development of standardised training plans for all training programmes.

TVTC's early activities involved setting up 10 new public/private partnership colleges in 2013 with a total capacity of 25,000 students.[14] This has necessitated a revision of National Occupational Skill Standards – a process that has been underway for some time – which for 10 years have been central to the management of TVET across the country and remain essential in standardising training throughout vocational institutions. By 2017, TVTC will have developed and launched 50 technical colleges, 50 girls' higher technical institutes and 180 industrial secondary institutes.[15]

The Saudi Petroleum Services Polytechnic

One of the flagship TVET initiatives within the oil and gas sector is the Saudi Petroleum Services Polytechnic (SPSP). SPSP is a vocational college that trains students looking to develop and further their career within the petroleum industry. Students are typically high-school graduates who have been employed

---

14. Technical Training College official sites.
15. UNEVOC Web site – http://www.unevoc.unesco.org/go.php?q=World+TVET+Database&ct=SAU.

within the industry and are sponsored by their employer to complete courses at the Polytechnic. The SPSP was established in order to provide relevant and industry-standard training solutions to the petroleum services sector and to be an institution that was able to adapt to the labour market and competency demands of the petroleum sector. The Polytechnic was established as a partnership between the Ministry of Petroleum and Mineral Resources, TVTC and Saudi Arabian Chevron. During the establishment of the SPSP, it was recognised by the Saudi partners that international expertise would be needed in order to provide the requisite range of vocational training options and support activities required to develop technically competent workers for the industry. To this end, the SPSP commissioned UK education services company TQ (part of the Pearson group) to operate the Polytechnic. Their activities include the provision of all academic and technical training, curriculum development and assessment. The Polytechnic offers qualifications that are internationally recognised by accreditation bodies as well as a comprehensive range of support services. In addition to the contract signed with TQ, SPSP reached agreement with Petrofac Training Services to operate and manage a construction, HSE and drilling training centre in Dammam.

The aim of the SPSP is to equip young Saudi Arabian nationals with the necessary skills and qualifications to make an immediate and valuable contribution to companies operating within the oil and gas sector. The Polytechnic offers a 2-year programme that starts with English language and also includes – alongside specific technical skills – mathematics, health and safety, employability skills and other relevant engineering tuition leading to the award of a Level 3 diploma.[16] TQ developed and now delivers the curriculum for all aspects of the training programmes offered and provides accreditation through two internationally recognised bodies: City & Guilds (a UK-based accreditation body) and the Process Awarding Authority/Vocational Qualifications–Science Engineering Technology (PAA/VQ-SET) – a specialist UK awarding organisation. The International Association of Drilling Contractors (IADC) is also due to become part of the accreditation offer.

The programmes offered at the Polytechnic enable students to achieve a number of qualifications including the following:

- Key skills certificate (which focuses on mathematics, health and safety and IT studies)
- SPSP English language certificate (offered at four levels and focused on English within a technical context)
- National Vocational Qualification (NVQ) Level 2 diploma in performing engineering operations (with students able to take one of six specific technical pathways – mechanical, drilling, electrical, operations, pipefitting and welding)

---

16. TQ Factsheet.

- Certificate in drilling services (which is accredited by the IADC)
- TVTC diploma

The SPSP campus at Dammam offers students access to five fully equipped workshops – electrical, mechanical, drilling and operations, pipefitting and welding. Academic training (English language, technical maths and health and safety) is predominantly delivered in classrooms. The campus also has a suite of language labs and is home to the Learning Resource Centre.

The imminent launch of the Construction and Drilling programme at the SPSP has necessitated the development of a new Construction and Drilling Training campus. Furthermore, the SPSP plans to build and manage a new polytechnic in the Al Khafji area in order to meet increasing demand for SPSP graduates.[17]

The success so far of the SPSP model can be attributed, in no small part, to the deep involvement of industry partners (with these organisations integral to the establishment of the Polytechnic as well as being responsible for sponsoring student through programmes). Furthermore, the adoption of international standards for certification (from UK bodies and from the IADC) strengthens the Polytechnic's claim to offer industry-standard qualifications that are internationally recognised.

The National Industrial Training Institute

In addition to the SPSP, TVTC has also been key partner in the establishment of the NITI. NITI is located in Al-Hasa and is built on land originally provided by Saudi Aramco. The facility offers industrial training predominantly to high-school graduates to qualify them to work in the oil and gas, petrochemical, energy and construction industries, as well as for the various services companies related to these industries. NITI opened its doors to students in 2014 and aims, through accredited programmes and on-the-job training, to offer accredited qualifications that are benchmarked to international standards. NITI is an independent, not-for-profit training institute and is a strategic partnership between Saudi Aramco and TVTC.

The NITI facility is designed to offer staff and trainees a fully functioning community within which they can live and work. The NITI complex provides accommodation for 1000 NITI trainees in fully equipped and furnished housing units. Accommodation is offered to any trainees who live more than 50 km from the NITI Complex, ensuring that the education and training on offer is available to the widest number of candidates, regardless of their geographical location. NITI staff are also accommodated inside the Complex.

The training programmes delivered at NITI have been developed under the guidance of the Technical Advisory Council (TAC), who are also responsible for providing recommendations for technical improvements across the institution. The TAC is made up of technical consultants who have been directly appointed

---

17. Information taken from the SPSP Web site.

by one of NITI's client organisations. The Council play an important role in supervising technical programmes by undertaking the following activities:

- Providing analysis of current and future industry requirements in terms of training needs
- Benchmarking NITI programmes against national and international best practices
- Providing recommendations for improvements to NITI's industrial and technical training programmes
- Establishing and building better channels of communication between NITI and its customers and partners
- Ensuring that NITI is making the best use of its network of founding members and other related organisations in relation to the ongoing funding of the NITI facility and the upgrading of technical equipment

The NITI facility houses a library as well as a number of simulators to support hands-on training activities to develop operational competencies. The facility also boasts virtual classrooms for e-learners and offers students access to a wide range of e-learning resources. Workshops have been installed to support competency development in a range of fields, including electrical, machinist, instrumentation, technician, construction and welding. The facility also offers candidates access to four science labs.

NITI offers a wide range of craft disciplines to candidates. These include the following:

- Instrument Repairman
- Telecommunication Technician
- Maintenance Electrician
- Electric System Operator
- Maintenance Machinist
- Pipefitter-Fabricator
- AC Serviceman
- Welder
- Heavy Equipment/Crane Operator Certification
- Lab Technician
- Critical Technical Programs: e.g. Renewable Energy
- Inspection Engineer
- Safety and Loss Prevention Technician

A key aspect of the NITI experience is the on-job training (OJT) component. This is particularly important to the Apprenticeship Programme that is run at the Institute. This Programme involves 2 years of study at the Institute followed by 6- to 12-month OJT designed to give candidates the technical skills and competencies they need in order to fulfil operational, maintenance, administrative and clerical jobs. Those candidates who undertake the Apprenticeship Programme are eligible to become full-time employees at their sponsoring companies if they successfully complete their training programme.

NITI also offers courses and short training programmes relating to a range of job requirements within the petroleum industry. These short courses are designed to provide professional training and competency development for particular disciplines. Many of these courses are offered through a dedicated e-learning portal and can be completed entirely online. Programmes offered include safety courses (Firefighting, First Aid, Chemical Hazards, DIP and H_2S), Helicopter Underwater Escape Training (HUET) and assimilation courses (Work Ethics, Safety, Wellness, Fitness and Citizenship Value).

The Colleges of Excellence

The Colleges of Excellence were established to be the principal institutions for delivered applied training across the Kingdom. The Colleges of Excellence represent one of the most significant manifestations of the TVET reform programme being implemented by TVTC.

Every College involves a collaboration with an international training provider. The Colleges of Excellence offer certificates and diplomas in specialised application areas for Saudi Arabian high-school graduates. During the first year of study, the Colleges of Excellence focus on developing foundation skills, including English language and communication skills.

Some facts about the Colleges of Excellence[18]:

- There are 10 colleges currently open with an additional 27 planned
- As of October 2014 there have been 140,000 applications for places
- Enrolled students currently number 9441
- Centres are managed by 14 international partners with more partnerships in the pipeline

Saudi Arabian oil pipes passing through Bahrain.

18. Figures taken from the Colleges of Excellence Web site.

Although none of the Colleges of Excellence currently offer specific courses in oil and gas disciplines, a number of colleges offer electrical and mechanical technology qualifications alongside training for the construction industry. This adds significantly to the pool of talent that the oil and gas industry can draw on.

THE IMPACT

The SPSP has made significant progress since launch. The success of the SPSP has been largely attributed to the level of industry involvement, the quality of the facilities on offer and the fact that candidates have already been offered employment by companies within the sector. Achievements to date include the following[19]:

- 100% of students have completed their Level 2 Specialised Award qualification
- Over 80% of students have completed their Level 3 City & Guilds Diploma, and obtained the equivalent TVTC Diploma. They have then progressed into successful employment.
- The curriculum has been developed in close collaboration with major employers in the Eastern Province of Saudi Arabia such as Saudi Aramco to fully meet their developing needs. This included involving employers on the board of trustees and developing the employers' existing material both to an NVQ framework and to meet international standards.
- The contract with the main supplier – TQ – was initially for a period of 5 years and has subsequently been extended up to December 2018.
- The success of the first Polytechnic in Dammam has led to the award of a second SPSP in Al Khafji with a capacity of 3000 students, representing a fourfold increase in capacity.

The NITI opened to students in 2014. At the time of writing, the first cohort of students has yet to complete their studies.

Aramco campus, The Leadership Center, Rahima, Saudi Arabia (Photo: Eugene Sergeev/ Shutterstock.com).

19. All figures taken from TQ Factsheet.

The Colleges of Excellence are growing at an unprecedented rate with 10 open as of October 2014 and a further 27 planned to open in the near future. The impact is difficult to measure at this stage with the initial tranche of Colleges opening in September 2013. Figures published by TVTC suggest that applications from Saudi Arabian nationals to study at the Colleges far outstrip the available places. This would suggest that the programme has been successful in selling the idea of a technical and vocational education to Saudis who, previously, would have been more likely to follow an academic pathway. That said, the true measure of impact will only be felt when the first graduates from the Colleges of Excellence graduate in 2017 and join the workforce.

THE CHALLENGES

It is difficult to assess the challenges associated with the TVET reform programme currently in process in Saudi Arabia. However, anecdotally, there have been a number of key messages that reflect the nature of the reform taking place:

- The pace and scale of change is bringing with it an inevitable challenge around quality assurance. The levels of investment currently going into the TVET reform programme are unprecedented. This means that there are a number of initiatives (three of which have been discussed here) that are being implemented simultaneously. This puts considerable pressure on the government authorities (and specifically on TVTC) to get things right.
- There is also a challenge around ensuring that the new institutions being established are able to attract and retain the best faculty, teachers and trainers from around the world. This is a critical aspect of any TVET effective solution. Currently the number and quality of suitably qualified and experienced teachers and trainers in Saudi Arabia who are able to deliver against the requirements of the expanding number of TVET institutes is not sufficient. These institutes can sometimes struggle to attract faculty from overseas and this can leave a shortfall in skilled staff. Staff attrition is also likely to be high now the institutes are beginning to run at full capacity.
- Saudi society still has a distance to travel in terms of changing the perception of technical and vocational education and attracting capable students into TVET institutions. The historic (and very common) perception that an academic education is superior to a vocational pathway will take some time to change, although early indications are that the new breed of internationally accredited institutions are growing in stature in the eyes of the Saudi Arabian population.

THE COST

It is impossible to estimate the overall costs of the Saudi government's investment in the TVET reform programmes that have been undertaken in recent years. However, there are some indications of the figures involved in specific aspects of the projects being undertaken:

- In June 2013, the Saudi government signed an SR 4 billion contract with leading colleges in the United States, Canada, the United Kingdom and Spain to provide technical training for Saudis. This equates to around $1 billion USD. These contracts were signed between TVTC and the international providers for the establishment of the first 10 Colleges of Excellence.[20]
- With TVTC planning to open up to 100 Colleges of Excellence over the coming years, the eventual investment in this programme alone could reach 10 billion USD.[21]
- In April 2014, it was announced that UK education providers won four contracts worth £850 million to establish 12 technical and vocational training colleges in Saudi Arabia.[22]

The Getenergy View

- The Saudi Arabian approach relies heavily on a web of public–private partnerships. This will require there to be strong leadership in order to manage the public–private partnership model and to ensure that the contracts being signed are delivered on in terms of impact and value for money.
- The motivation of corporate entities and the motivation of TVTC may clash. Private companies wish to extend their contracts and this may not improve the TVET system or benefit the education system as a whole. If you privatise the education system, internationalise the faculty and create a barrier to the sense of identity, and work on 3- to 5-year contracts, this may not produce the large numbers of field-ready professionals needed across the industry and employability may be questionable.
- This is a highly commercialised (and effectively privatised) system of TVET. Although this allows for the involvement of international providers, it changes the context for TVET in Saudi Arabia significantly. Are private sector organisations as invested in a successful end result as if the education system had been built from within?
- There are questions over whether the Saudi Arabian government has the requisite experience to undertake outsourcing at this scale. With billions of dollars being spent with overseas companies, the governance of these contracts is extraordinarily challenging.
- There is a danger that those international education and training companies who are winning contracts may find the operating environment rather more difficult than they expect. The level of investment may be high, but there are

Continued

---

20. Arab News, 13th June 2013.
21. Arab News, 2014. *High-tech studies: Kingdom setting world standards.*
22. Government Press Release, 2014. *£1 billion exports win for UK education in Saudi Arabia.*

The Getenergy View—cont'd

already questions over the profits of those companies involved and this calls into question the long-term sustainability of the approach.

- Although the scale of the reform programme is to be commended – and the attempt to 'rebrand' technical and vocational education as a valuable education is admirable – there are doubts over whether the size and pace of change may inevitably lead to challenges around the impact and quality of what is on offer.
- There are real challenges in terms of fulfilling teaching and training positions with evidence of poaching within the region and a paucity of suitably skilled staff. There is a real need to instigate significant train-the-trainer programmes in order to create a locally based skilled cadre of teachers and trainers.

A Note on Sustainability

The level of investment going into the reform programme is vast and although this reflects the strength of the Saudi Arabian economy, there are clear questions around how long this level of investment can be sustained. More pertinently, the degree to which interventions are successful will ultimately define whether the approach is sustainable and it will be some time before that is clear.

A Note on Replicability

Many of the aspects of the approach being taken in Saudi Arabia would be replicable within another context. In fact, the institutional approach to TVET reform and the focus on industry-specific colleges is something that has been seen in a number of other countries. Whether many other countries would be able to meet the level of investment currently being ploughed into the system in Saudi Arabia is an entirely different question.

A Note on Impact

On one level, the impact of the reform programme has already been profound: the number of new colleges opening up (and the speed at which they are opening) is creating a new landscape of technical and vocational education, one driven by international partnerships. However, the real impact will only be known when graduates of this new system enter the workplace.

Case Study 3

Mozambique: The Birth of a New Energy Nation

How a Future Energy Powerhouse is Planning to Address the Need for Competent, Locally-Trained Employees

Chapter Outline

This case study was heavily informed by a Getenergy workshop held in Maputo, Mozambique in February 2014 involving the energy industry, education and training providers and representatives from the Mozambique government. The authors would like to thank all of those who took part in this workshop.

THE MOTIVATION

Mozambique has a long history of hydrocarbon production. However, the recent discovery of a gas field off the northern coast has the potential to transform the fortunes of one of the poorest countries in Africa. In 2012, exploration activities by the US oil group Anadarko discovered an estimated 850 billion cubic metres of natural gas in the Rovuma basin[1] – more than three times the reserves left in the North Sea. Soon after, the Italian energy group Ente Nazionale Idrocarburi (ENI) also made two significant discoveries nearby.

Mozambique now stands at a crossroads. The economic and societal benefits that ownership of major hydrocarbon reserves can bring will only trickle

---

1. Geological Survey, December 2013. *Minerals Yearbook: Volume 3: Area Reports: International Review: 2011, Africa and the Middle East.*

Education and Training for the Oil and Gas Industry: Building A Technically Competent Workforce.
http://dx.doi.org/10.1016/B978-0-12-800975-8.00003-4

down to the general population if the country is able to build an energy sector strategically and ally this to proactive skills development and labour market policies and practices. The threat of the 'resource curse' looms large. While there are many international organisations – the World Bank, the International Monetary Fund and others – who are working to ensure Mozambique succeeds in achieving sustainable and equitable economic growth, the process of building an energy industry – and ensuring that the population as a whole benefits from this – will take many years and will require significant political will and vision.

Current levels of inward investment into Mozambique demonstrate that the energy industry is here to stay. The available reserves mean that Mozambique has the potential to become one of the wealthiest countries in Africa. In a global energy market where hydrocarbons are increasingly hard to find, the offshore discoveries have already attracted many of the world's leading energy companies to Mozambique. However, the country currently has virtually no education and training infrastructure to support the development of nationals as competent employees and technicians for the industry. The wider education and training system is not equipped to meet the demands of a twenty-first-century economy and the education and employment prospects for a majority of young people are bleak. Without the means to develop the skills and competences needed for the fast-growing domestic economy, the economic and employment benefits that the energy industry (and related sectors) can offer – both during the exploration and construction phase and beyond – will not be felt by those who most need it, namely the poor, unemployed and underemployed of Mozambique. The historic pattern seen in a number of other countries with hydrocarbon reserves will be repeated – available jobs will be fulfilled by an expatriate workforce and benefits to the local population will be minimal.

Mozambique now faces both a challenge and an opportunity. The challenge is to capitalise on economic and industrial expansion driven by the energy sector without simply selling the industry to international operators and succumbing to the temptations of quick money with no thought for community or societal legacy. The opportunity is to use the revenues and inward investment into the energy industry to build a new education and training infrastructure that can do two things. First, it can supply the emergent energy industry (and many of the related service industries) with well-qualified local candidates who will be able to competently do the job, compete for employment opportunities and earn a living wage. Second, the opportunity is to create an educational legacy that will serve the workforce requirements and skills development needs of the country beyond the energy industry, giving citizens access to training and development and improving the talent pool for the country as a whole. Against this backdrop, the motivation to get this right is significant and the potential rewards for the country as a whole vast.

THE CONTEXT

According to the World Bank, only 12% of Mozambique's population have access to electricity and the country is one of the poorest and least developed nations in the world. At the same time, economic development is rapid with average annual GDP growth rates at around 7.2% for the last decade.[2] However, this growth has been driven by a boom in the capital-intensive extractive industries and, subsequently, has failed to generate significant employment. As a consequence, very little of the wealth generated by steady economic growth has reached the broader population.

The country is now in the midst of a rapid industrial expansion programme that is driven by the coal, gas and construction sectors. Mozambique is projected to become the world's biggest coal exporter within the next decade[3] and the recent discovery of two massive offshore gas fields has marked the country out as a major area for energy investment globally. Some estimates put the revenue potential for the nascent liquefied natural gas (LNG) industry at around $420 billion. Investment is pouring in from overseas – from the US, China, Europe and beyond – driven in part by the political stabilisation that has taken place as the country emerged from a traumatic civil war. According to Reuters, Mozambique's was the best performing currency in the world against the US dollar in 2011.[4]

Expectations for Mozambique are high. As business and industrial activity grows and the middle class expands, the country is beginning to see improvements in the level of professionalism and service delivery across a range of sectors. The capital city Maputo is rapidly becoming a modern metropolis and investments in construction and infrastructure are significant.

The port in Maputo, Mozambique.

The tension between Mozambique's relative poverty and expanding economy – and the potential that recent hydrocarbon discoveries offer – has attracted the interest of international donors with these organisations keen to

2. Africa Economic Outlook, Mozambique 2012.
3. KPMG, Mozambique Country Mining Guide 2013.
4. Consultancy Africa Intelligence, 2012. *The "Resource Curse" in Mozambique.*

guide the country in a positive direction and support the sustainable growth of the economy over the coming years. The timescales for development are the key – it is likely to take between 5 and 7 years for the country to be in a position to export any of the newly discovered natural gas and require tens of billions of dollars of investment in order to get to that point. A long journey lies ahead.

Education in Mozambique

If Mozambique is to use the emergent LNG industry as a catalyst to stimulate employment and drive skills development, there will need to be significant investment in, and radical change at, every level of education. The country faces a number of profound challenges in regard to education and training:

- Mozambique has the lowest level of average schooling per capita in the world with students spending an average of 1.2 years in education.[5]
- Enrolment into both secondary and tertiary education is much lower in Mozambique than in neighbouring countries. Financial constraints drive very high dropout rates, with most families unable to afford to keep learners in education or training.
- The poor level of relevant skills within the labour force remains a significant issue for employers looking to recruit locally. As a consequence, many companies import labour in order to meet their needs, with Portugal providing significant expatriate labour due to the lack of employment opportunities there. The shortage of skilled local workers in the extractive industries is particularly acute.
- There are significant challenges around the quality and relevance of technical and vocational education, with the system fragmented and little engagement between educators and employers.
- There is a need to change the culture of tertiary education and to put in place a system that can develop a skilled workforce that is better qualified, is more competent and creates competitiveness within the job market; there is also a need to cultivate an entrepreneurial spirit among citizens.
- There is currently a lack of clear pathways for the graduates of vocational education to progress into available employment opportunities.
- Part of the process of improvement sits with training institutions – they need to be able to provide relevant services and high-quality training in an efficient and financially sustainable manner supported by up-to-date facilities, sound pedagogy and modern industry-standard curricula.

Although the ambition for skills development may be clearly articulated by the Mozambican government, the solution will need to encompass every aspect

---

5. United Nation, Human Development Report 2013. *The Rise of the South: Human Progress in a Diverse World.*

of the education and training ecosystem and, within this context, it must be recognised that progress will be incremental and investment considerable.

Employment in Mozambique

The potential that the extractive industries have for transforming the national employment landscape is significant. The current context illuminates some of the challenges that lie ahead:

- According to a labour survey from 2004–2005, 75% of working Mozambicans were self-employed, with a majority working in the agriculture sector.[6] Of those who were in work, many supplemented their income with secondary economic activities due to the fact that one job did not provide them with a living wage.
- Men and women have very similar employment rates (unlike many other countries in the region), reflecting a degree of equality between genders in Mozambique.
- The Benga coal mine in Tete Province is set to become one of the largest coal mining projects in the world. However, it has only created 150 direct jobs, with a projected 4500 further jobs set to be created in the near future.[7] This gives an indication of the negligible impact on employment that expansion in the extractive industries can have.
- Urban youth unemployment is a significant problem, with the formal employment sector not able to absorb the estimated 300,000 young people entering the labour market each year.[8] This means that the informal economy thrives and many young people end up in poorly paid jobs with little or no job security.
- It is estimated that only 5.1% of the total labour force is in formal employment.[9] Most informal employment is in subsistence farming. Outside agriculture, the informal sector is still the largest employer, with growth estimated at 7%–8% per year.[10] There are an estimated 700,000 jobs in the formal sector.[11]
- Agriculture is by far the single largest employment sector in Mozambique, employing around 80% of the labour force, though only contributing 31% of GDP.[12]
- Despite massive unemployment and a lack of opportunities for living wage jobs in the formal sector, there are significant numbers of foreign nationals working in Mozambique. Foreign companies are often accused of breaking

---

6. Africa Economic Outlook, Mozambique 2012.
7. The World Bank, *World Development Report 2013: Jobs*.
8. Ulandssekretariatet, LO/FTF Council. *Mozambique – Labour Market Profile 2012*.
9. Ibid.
10. Ibid.
11. Ibid.
12. Ibid.

local labour laws and quotas for foreign nationals in firms have now been established.

- The national energy company of Mozambique, Empresa Nacional de Hidrocarbonetos (ENH), estimates that 16,000 skilled individuals will be needed in the energy industry by 2020 (with this figure reflecting demand during the construction phase). During the operational phase, around 2500 direct employees will be required, although many more will be needed to support related sectors.

Part of the challenge facing Mozambique is to connect demand in the extractive industries (and in the related sectors that will grow as a result) to the supply of skilled workers. At present, there is little connection between the two, leading to skills shortages within a context of high unemployment.

The Extractive Industries

Mozambique has experienced a period of sustained economic growth since early in the twenty-first century. This has been driven predominantly by the extractive industries. With the major natural gas discoveries of 2012, this is set to continue:

- Mozambique holds an estimated 4.5 trillion cubic feet (Tcf) of proven natural gas reserves, but does not have any crude oil reserves as of 2013, according to the *Oil and Gas Journal*.[13]
- Mozambique has large onshore and offshore sedimentary basins that contain hydrocarbon resources. Hydrocarbon exploration began in Mozambique in 1948. Today, according to the Petroleum Institute of Mozambique, the country has four active gas fields, all located onshore in the Mozambique basin at Pande, Buzi, Temane and Inhassoro.
- In 2011, Mozambique produced 135 billion cubic feet (Bcf) of dry natural gas from the Pande and Temane fields, which are both operated by Sasol, a South African energy and chemicals company.[14] Most of the dry natural gas produced in Mozambique was exported (117 Bcf) to South Africa via the 535-mile Sasol Petroleum International Gas Pipeline and the remainder was domestically consumed (18 Bcf).[15]
- According to a 2012 assessment commissioned by the World Bank, Mozambique contains almost 46.7 billion barrels of oil equivalent, or 279.9 Tcf equivalent, of discovered and undiscovered oil and gas resources. More than 70% of these resources are in the Rovuma offshore north region.
- There have been a series of natural gas discoveries in the offshore Rovuma Basin that are large enough to support LNG projects. Anadarko and Eni have

---

13. EIA – http://www.eia.gov/countries/country-data.cfm?fips=mz.
14. EIA – http://www.eia.gov/countries/regions-topics.cfm?fips=eeae.
15. Ibid.

led exploration activities in the area and Mozambique expects to begin exporting LNG to the global market by 2020.

- Anadarko and Eni may also build separate LNG facilities to utilise the natural gas produced in the Golfinho/Atum and Coral areas. Eni have submitted plans for a floating LNG facility in one of the exploration areas.

- Direct economic benefits from the gas industry are likely to be significant, with some estimates suggesting revenues of between $200 billion to $400 billion over 40 years.[16] In a country where per capita GDP stands at $1100 and government spending is at $6 billion per annum,[17] this level of revenue will have a transformative effect.

- In 2014, the government of Mozambique, through the Ministry of Mineral Resources (MIREM), published details of several concession contracts for exploration in the petroleum and mining sectors. This demonstrates the broader commitment to expansion of the extractive industries to include oil, heavy sands, coal and other minerals.

- Tete Province boasts one of the largest coking coal mines in the world. The giant Brazilian mining firm Vale is spending billions on operations including the construction of a coal terminal and railways. It aims to reach a production capacity of 22 million tonnes by 2014.[18]

Within this context, the task ahead for the Mozambique government – and for others engaged in industry and the expansion of education and training – is daunting. Economic and industrial growth is already driving up demand for skills, generating employment opportunities and creating heightened expectations among citizens and communities. The challenge now is to leverage inward investment and capitalise on the interest and engagement of others to put in place a strategy that will achieve the impact that so many watching believe is possible.

THE SOLUTION

The solution to Mozambique's skills and workforce development challenges is multifaceted and will require the involvement of many organisations within the country and beyond. The approach adopted by the government of Mozambique has so far been informed by extensive engagement both nationally and internationally. The evidence suggests that their activities have been underpinned by a number of key principals:

- That a strategic approach is necessary in order to develop the capacity of Mozambique's education and training system to meet the skills requirements of the oil and gas industry and to invest in national development thereafter.

---

16. Mozambique Background???
17. Ibid.
18. Vale (2010). "Expansion of Moatize, Mozambique." Retrieved February 2, 2012 from Vale.

- That this strategy requires the closest possible partnership between government, the oil and gas companies and the education and training institutions, and that structures must be established to ensure these partnerships work effectively.
- That measures are required to develop mutual trust, understanding and collaboration between the three partner groups (government, the oil and gas companies and the education and training institutions) as a basis for current and future collaboration.
- That strong international links and collaborative projects are required to help develop and implement the right education and training programmes and qualification systems that will ensure the maximum participation by Mozambicans in forthcoming projects in the extractive industries.

The actions taken at a local, national and international level must be informed by a clear and coherent vision. That vision is evolving, but current discussions involve the following aspects:

The education and training system should reflect and aspire to the following ambitions:

- A holistic approach to education and development – funded in part by hydrocarbon revenues – that encompasses areas like community health and nutrition
- The implementation of education and training programmes that support workforce and skills development for the industry and that are in tune with industry needs and requirements
- The development – over time – of an education and training infrastructure that supports workforce and skills development for the long term: this means not only a focus on the short-term requirements of the extractive industries but a recognition that the energy sector will never provide employment for the whole population and other skills and competencies need to be nurtured
- Support for the development of a pool of talented and educated people across the whole industrial landscape and in many different disciplines, including construction, upstream, natural gas-based industrial development and so on
- Significant improvements to education at every level with a particular focus on improving basic skills, increasing access to education (particularly in rural areas) and improving education outcomes: the proceeds from the energy industry should be utilised effectively to make the necessary investments in teachers, facilities and the wider education system
- A vocational education system that embraces practical, hands-on learning, has better qualified technical training staff and greater availability of 'train-the-trainer' programmes
- A vocational and technical education system that is underpinned by public–private partnerships (PPPs) for funding
- A long-term commitment from all participants to a process of education reform that will take decades to fully bear fruit on society, industry and community

- An education system that gives young people the capacity to make informed choices about their education and career and that exposes them to the opportunities within the energy industry in Mozambique and beyond
- A smooth transition from secondary education to technical education with candidates from technical education well-prepared so that they can learn and attain a requisite level of competency
- Vocational institutions that meet regional needs so that available courses are matched with what the industry requires in that specific region
- An effective education system that is able to recognise pre-existing vocational skills (recognition of prior learning) and that can certify and establish competency among the existing workforce
- A continuous dialogue between industry and training institutes to design curricula and fund training programmes
- An awareness of global case studies within education and training and an ability to learn from and implement international success stories
- An open market for non-state skills providers
- An established coordination mechanism to identify skills gaps and prioritise and fill these gaps
- A cadre of highly qualified local instructors and coaches who are able to transfer knowledge and skills to others over time and the facilities in place to train and develop these instructors
- A system of financial incentives in place to reward good work in the area of technical and vocational education and training and that provides bursaries for local students

The vision for the energy industry is as follows:

- A nationalised workforce for the energy industry and supply chain where opportunities for young people, existing workers, entrepreneurs and others are supported and facilitated by access to relevant and high-quality education and training
- A range of international partnerships that work for Mozambique and that support skills development, the growth of key industries, the long-term sustainability of education and training and the wellbeing of communities: this will include partnerships with governments, development organisations, educational institutions, private sector companies and others
- An energy industry in Mozambique that is safe, well managed and efficient and that is a model for how other countries can grow and manage their hydrocarbon sectors
- An energy industry that is able to engage with and use non-national workers and companies effectively to expand operations and make them cost-effective, but that does not become overly reliant on this imported workforce at the expense of nationals
- A stable legal and contractual framework – developed and implemented by the government – to enable the energy industry to grow in a responsible and ethical way

- A regulatory framework that ensures local Mozambicans fill the skills gaps in industry and that helps the country avoid the 'resource curse', with strategies to increase the number of qualified locals and reduce a reliance on foreign workers
- A coherent, clearly defined plan for workforce development and an education and training strategy that involves and engages with all key organisations, including oil and gas companies, service companies, engineering, procurement and construction companies, government bodies, education institutions and others
- An energy industry that has a critical mass of talent from which it can choose and that meets the demands of the industry across different disciplines

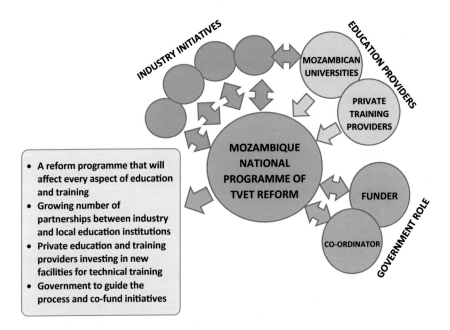

The impact on the Mozambican economy and the wider society should encompass the following:

- Communities – particularly those in areas where hydrocarbon extraction is taking place – that see employment and economic benefits from the emergent LNG industry and that are fully engaged in the decisions that affect them
- A clear understanding of the workforce requirements for the energy industry and beyond and a mechanism to track and monitor those requirements over time – a move to a demand-driven rather than a supply-driven system

- A means of addressing the disparity between the skills base within the informal economy and the requirements of the formal sector and a mechanism to transition individuals between one and the other
- Adequate and effective support for the development and sustenance of small- and medium-sized enterprises (SME) in Mozambique with a focus on enabling new and existing SMEs to gain access to the opportunities there will be to supply goods and services to the extractive industries and the conditions to help develop and nurture local entrepreneurs who can be part of the supply chain
- Greater investment in local communities that focuses both on immediate needs – clean water and education – and on the type of regional and national investments that look at longer-term challenges
- A process for the promotion of PPPs that ensures these partnerships achieve impact and value for money
- Individuals within the workforce that have access to continuous professional development so that they can be adaptable and evolve along with the projects

In order to achieve the vision, a number of actions will need to be undertaken both immediately and over time. The draft plan that emerged from engagement with all key parties summarises these steps:

Immediate Actions

- Establish a set of advisory groups that will involve participants from the government, education and training institutions and oil and gas companies. If these collective interests are taken into account, then the chances of successful programmes and projects being initiated will increase. Initially, three groups should be established – one for higher education, one for technical and vocational education and training and one for school-level education.
- Survey and describe all current projects concerned with workforce and skills development for the industry. This will give a clear picture of the situation in regard to the supply (and anticipated supply) of skilled employees into the workforce over the coming years.
- Build – in collaboration will all industry partners – a more detailed picture of the direct and indirect workforce demands emerging from the nascent energy industry. Some work has begun in this area, but more collaboration from industry is required and the government of Mozambique needs to be more active in exploring the wider employment and workforce implications of the emergent energy sector.
- Create a strategic plan for workforce development based on a comparison of the supply and demand to establish what is needed on the supply side to meet the demand. The strategy should build on what is already being done and also identify additional education and training interventions that will be needed now and in the future.

- Identify the partnerships – national and international – that will be required in order to deliver against the strategic plan and begin a process of engagement and consultation with relevant partners to move this forward.
- Facilitate engagement between the industry and existing institutions in Mozambique to establish approaches and mechanisms for raising standards to acceptable levels and implementing training that is industry-approved.
- Identify ways that, in the short term, the international community can collaborate with Mozambique around the oil and gas industry in the form of scholarships so that the current shortfall in local education capacity can be met, in part, by placing individuals into relevant institutions internationally.

Future Actions

- Ensure that government and industry work together to integrate and adapt international standards within the energy industry and identify what is applicable in Mozambique in order to have unified standards across the board.
- Establish partnerships between education institutions in Mozambique and international institutions to build knowledge and capacity and to promote research.
- Create a formal system for collaboration between education and industry in Mozambique that sets the parameters, provides guidelines and ensures efficacy and impact.
- Establish labour market observatories that are able to measure and track skills requirements and is charged with sharing this information with industry, the government and education providers.
- Develop a master plan for human resource development that encompasses the requirements of the energy industry but that also looks at workforce development for every sector and every part of the country.

There are, at the time of writing, a number of initiatives being undertaken by the government and associated governmental agencies that indicate the direction of travel. These are summarised as follows:

The Government of Mozambique

- The government is engaged in an active 5-year programme where the main objective is to fight poverty and improve the life of Mozambique people. The scope of this programme must include activities in the extractive industries due to the importance to the natural gas and heavy sands reserves.

- The government of Mozambique approved (in 2013) policies for extractive industries that establish how minerals will be produced. These policies look at links with other industrial sectors and at the promotion of skills.
- The government also approved (in 2014) a policy of social responsibility and developed a framework for the implementation of social responsibility that supports organisations to develop local industrial plans and activities. The promotion of local economic development and the capacity of local Mozambicans is part of this approach.

National Institute of Employment and Vocational Training (INEFP), Ministry of Labour

- INEFP recognise that there is a gap between labour demand and the capacity of education and training institutions to respond to this demand. To address this, they are tackling three main issues:

  - *Infrastructure and training facilities/vocational training institutions*: there are, as of 2014, 15 vocational centres nationally; the aim is to double this number by 2020. Young people in rural areas cannot currently access vocational training, so INEFP are establishing mobile units for training. The plan is to establish 33 mobile units, 3 for each province.
  - *The number of qualified trainers*: there are, according to INEFP, less than 250 qualified vocational trainers currently working across the country; the aspiration is to have 3000 active trainers and to improve competency among existing trainers.
  - *Qualifications offered in the centres*: the current offer does not respond or map to the profile of competencies needed by industry; there is a need to understand the industry qualifications and skills that are required, and INEFP are working with the International Labour Organisation and Sasol to better understand the requirements.

The programme of vocational education reform (PIREP) and the Ministry of Education

- PIREP was established to facilitate the transition from the current system of vocational education to a demand-driven system of training that is focused on the development of industry-relevant skills and the promotion of employment opportunities within high-growth sectors.
- The programme seeks to ensure that graduates of vocational education are better able to access opportunities in the labour market and that the companies and organisations where they work have improved efficiency and competitiveness.

- Their work is also focused on helping training institutions to develop and provide relevant services and quality training in an efficient and financially sustainable manner through a decentralised, competent and credible approach with clear links to the local labour market.
- The Ministry of Education are actively engaging the production sector in curriculum design and companies are participating in validation panels that develop professional courses (including in gas and mining).
- PIREP are also actively looking at programmes that will train the trainers and are engaged in a process of re-equipping schools.
- The Ministry of Education have worked with the Scottish Qualifications Authority (SQA) on the development of competency standards.
- They are also in the process of developing a national qualifications framework and have recently created a new directorate for qualifications.

THE IMPACT

The implementation of the government's strategy is in the early stages. It will be some years before the impact is truly measurable. However, there are a number of initiatives being run locally and with international support across the country to address the development of competent, suitably skilled workers for the energy industry and that reflect the ambitions of those involved to achieve the vision set out earlier:

Eduardo Mondlane University (UEM)–Sasol

- In 2014, UEM signed a memorandum of understanding with Sasol to provide a framework to develop capacity building projects.
- The aim is to improve the quality of undergraduate training and grow academic networks.
- Activities will also focus on improving the capacity of teaching staff, offering scholarships for students, establishing a downstream curriculum, building a laboratory with world-class equipment to give students hands-on experiences, designing new and relevant curricula, encouraging better teaching methodologies and improving teaching facilities.

Anadarko–UEM–Texas A&M University

- Anadarko Mozambique Area 1 recently launched a petroleum engineering programme with UEM and Texas A&M University (and involving a number of other US universities). The programme aims to develop the petroleum engineering graduates for the future of the energy industry in Mozambique.
- UEM and Anadarko began their partnership in Mozambique in 2012.

- Texas A&M University provide the tools and resources that UEM, the faculty and students need in order to build a sustainable programme in Mozambique. Their focus is on creating a replicable, high-quality learning experience for the students.
- They currently use virtual capabilities to deliver parts of the programme with US professors teaching live via the Internet. The approach involves US professors co-instructing with their Mozambique counterparts on the ground. Eventually, 100% of instruction will transition to the Mozambique teaching faculty.
- The faculty at UEM is expected to grow – Anadarko are actively seeking partners to support this process.

ENH

- ENH are the national energy company of Mozambique. They have been engaged for some time in capacity-building projects for the industry that include the sponsoring of secondments to other countries.
- Their vocational training activities encompass 217 training programmes. The main focus areas for these courses are exploration, engineering, procurement, HR, finance, protocols and English language training.
- ENH are currently training senior staff at Galp in Portugal and at a training institute in Japan. They are also running training programmes in Italy and have a scholarship programme in France.
- ENH, with other international oil companies, are part of the consortium that is developing a large technical training centre project in Pemba that will support the nascent LNG industry offshore.

ENI

- Italian energy giant ENI are running school and student programmes in Pemba and they plan to replicate this in Maputo. Evidence suggests that in the Cabo Delgado region there is reluctance among the local population to leave and travel to Maputo, so training and education needs to be delivered locally.
- ENI are working to create a bridge between secondary school and university and to create an integrated approach to develop the talent pipeline. To this end, they have supported an initiative to engage young people in science, technology, engineering and mathematics (STEM) through exposing students to lecturers from ENI and University College London.
- They have offered to 150 graduates from Mozambique an intense 12- to 18-month programme in Italy to get them ready for the industry.
- ENI also recognise the fact that universities need the knowledge and expertise to teach and train individuals for this new LNG market. They are investing in the training and development of new professors in gas engineering who can be the faculty for the future of the industry in Mozambique.

Association of Canadian Community Colleges (ACCC)

- A project to develop skills in Mozambique has had an investment of $18.5 million over 6 years.
- The ACCC are working with local partners to build centres of excellence dedicated to developing skills for the natural gas industry in Cabo Delgado.
- This initiative could benefit as many as 4000 Mozambicans within the extractive industries and will include a focus on employability and on developing entrepreneurial and soft skills.

ISQ

- Portuguese education company ISQ has been contracted to develop the Pemba Technological School for ENH – this will contribute to the training and development of individuals for the construction, operation and maintenance of LNG plants in Mozambique. All courses will include English language, communication, health and IT skills.
- There will be 100 students annually completing courses with maximum of 400 students undertaking studies at any one time. The fields of study will include mechanics, construction, electricity, instrumentation and welding. Work-based learning will be an essential part of the approach.
- Graduating students will receive an international European diploma. The School will also be active in promoting the placement of graduates into the labour market.

THE CHALLENGES

As set out earlier in this case study, Mozambique is facing significant challenges in achieving its vision of a thriving economy that drives up living standards and shares the benefits of economic growth with the population. These challenges can be summarised as follows:

- *The need for patience*: building an education and training system that can truly support skills development at the level required in Mozambique will take time. Set against that are the short-term commercial demands of international oil companies and the need for the government to start capitalising on the hydrocarbon dividend as soon as possible. These competing timescales are difficult to marry.
- *Transparency is critical*: some local news reports have already hinted at a lack of openness in regard of the deals that the government is currently signing with international operators. It is vital that political leadership is shown in this regard and that the door is not left open to corruption.
- *It is about more than gas*: although the growth of the extractive industries offers real opportunity to many citizens, the benefits of the emergent energy sector will be seen if the government is able to direct investment in the right

way – this means spending on infrastructure, improving all aspects of education, investing in agricultural productivity and ensuring that the economy as a whole feels the benefits of LNG revenues.

- *Long-term thinking is required*: many of the projects and initiatives outlined in this case study demonstrate that the immediate focus for the government – and for other partners – is to develop mechanisms to nationalise the workforce within the energy industry. However, longer term, the number of jobs in the industry will diminish and, ultimately, the gas will run out. There needs to be a continued focus on a broader skills development agenda that will ensure other industries can grow over time and that citizens are given the opportunity to fulfil positions within these industries.

The broader challenge is simply the distance that Mozambique has to travel. As one of the poorest nations in the world, the transformation to a well-developed emerging economy can seem like a distant dream, although there is no doubt that such a transformation is entirely feasible.

THE COST

One of the significant challenges facing the Mozambican government is the level of capital investment required to meet the countries strategic objectives in terms of education and training infrastructure and development. This has to take place some years prior to energy revenues coming on stream. Meeting the short- to medium-term costs of developing adequate education and training provision across the country will only be achieved through the proactive participation of international partners.

The Getenergy View

- The scale of the challenge facing Mozambique is vast. Although that should not deter those who are trying to make a difference, there is a danger that amid the flurry of initiatives, the process of reform will become overly complex – if there is too much going at one time, control may be lost over individual projects. Progress needs to be strategically guided and incremental.
- Whatever education and training solutions are developed with hydrocarbon money must be able to withstand changes in government and financial fluctuations (particularly in relation to the price of gas). This will require clear leadership to hold together groups of stakeholders.
- At present, the strategy outlined in this case study is the result of initial discussions held in Mozambique between the government, the industry and selected education and training providers. In absence of clear identification of national priorities, there is a danger that the reform process will lack coherence.
- The language barrier is a real challenge in Mozambique – foreign companies will typically speak English and many training programmes are written and delivered in English. However, English is not the principal language that is

Continued

The Getenergy View—cont'd

taught in schools (Portuguese is) and so candidates for education and training and those looking to work in the industry will need significant support in bringing their English-language levels up to speed.

- The Mozambican government will need to demonstrate strong leadership in order to manage international participation in the ongoing reform of the education system. There needs to be a clear understanding of the role of private sector in developing national education.
- The wider economic objective should be to build an economy that can benefit from cheap energy, not just build an export economy that will be reliant on maintenance of the international oil price. In this way, Mozambique avoids 'resource curse' and diversifies its economy by enabling it to produce products, not just natural resources.
- The expectations on international oil companies – who are leading the early exploration activities in the country – are high in terms of their contribution to the development of the education and training ecosystem. There is a vast amount of infrastructure to build, but this must be balanced with the immediate business goals of these companies for the region.
- There is clearly pressure on the government to produce results in the wake of the massive gas discoveries made and the impact that this might have on the country as a whole. However, this must be balanced with the need to spend the correct amount of time to build the necessary infrastructure required for the proper development of oil and gas reserves.
- The government needs to take a hold of the process and lead the initiatives – the reform programme needs a framework to ensure the various elements fit together and that any efforts to develop petroleum engineers and geoscientists are distinct from general educational reforms.

A Note on Sustainability

The level of investment required – from the government, private sector and international donors – is vast. However, the sustainability of this investment is underwritten by the quantity of hydrocarbons in the ground and the prevailing oil and gas price. The challenge will come in a decade when investment falls off and new models of education and training have to offer greater financial viability.

A Note on Replicability

In some ways, the situation in Mozambique is unique. This is a country that is having to move very far and very fast. Furthermore, the reform programme is in the very earliest of stages with many years of change ahead. There is currently no other country on the earth that is facing the particular challenges that Mozambique is facing. How effectively they deal with these challenges – and the degree to which their response is replicable – will only emerge in time.

A Note on Impact

Many of the early initiatives that are now taking shape are likely to be effective in that they replicate models elsewhere. There is, at present, a significant focus on higher education and less impetus around technical and vocational education. For education reform to reach the widest number of people, more effort will need to be targeted at broadening and improving the TVET system, as this is where major impact across society will be felt.

Case Study 4

Houston: Building on Success

How Community Colleges in Houston Are
Addressing Workforce Development within
the World's Most Successful Energy Sector

Chapter Outline

With thanks to Debi Jordan, Executive Director of the Centre for Workforce and Community Development, Lee College

THE MOTIVATION

Over the last century, Houston has developed from a region largely dependent on agriculture to being the world's energy capital. Its history and changing economic landscape can be told through the story of three important oil 'booms': firstly, the boom that propelled Texas to become the leading producer of oil at the turn of the twentieth century; secondly, the series of oil and gas shortages, price hikes and then price plummets in the wake of surplus production in the 1970s and 1980s; and thirdly, the shale gas revolution that the state is currently witnessing.

The community college system was, and remains, an important factor in educating and preparing a field-ready workforce for the oil and gas industry throughout Texas. In the course of oil booms and busts, economic changes and the global financial crisis, Houston has emerged as an economic success

Education and Training for the Oil and Gas Industry: Building A Technically Competent Workforce.
http://dx.doi.org/10.1016/B978-0-12-800975-8.00004-6

story fuelled by oil and gas. Amidst wide-scale economic development, urbanisation, industrialisation and economic diversification, the education system has, and continues to, respond and contribute to fulfilling the needs of the oil and gas industry as well as responding to wider industrial activity and shifts in the social paradigm of the city – whether it be incorporating women into the local workforce, educating immigrants, or retraining veterans for employment.

However, success brings fresh challenges. While community colleges have played a vital role in supplying field-ready, technically competent workers, there continue to be new demands on the vocational education system. These demands require new initiatives, forcing a rethink in the approach to vocational education. These challenges are related to new developments within the petroleum industry across the Texas gulf coast region and within Houston:

- Over 120 petrochemical projects have been announced – with an estimated $80 billion (US) investment for the US Gulf Coast region.[1]
- In Houston, Greater Houston Partnership (GHP) has reported 41 projects that will provide investments totalling an estimated $18 billion in the run up to 2020.[2]
- The retirement of technically skilled professionals has left Houston with a deficit of experience and expertise.
- Estimates suggest that there will be more than 74,000 openings in middle-skills occupations in Houston between 2014 and 2017.[3]

Recently, concern has arisen as to how the region can train and develop its workforce in the face of these pressures. Indicators show that there is a worrying gap in the supply of middle-skills professionals in Houston's workforce.[4] Middle-skill jobs are generally defined as occupations requiring some level of higher education, either at the community college level or university level, and involving technical training.

This case will explore the historical effectiveness of the community college system across Texas, how it has responded to the industrial needs of Houston's rapid development as an energy hub and how this system will continue to achieve this aim in the future. The historical aspect of Houston's 'booms' and 'busts' are integral to telling the story of regional success – the ability to move away from a resource-producing economy and into a diversified, knowledge-based economy is rooted in vocational education and training. Houston's economic future is inextricably entwined with its community colleges and their ability to involve industry at the heart of the vocational

1. Greater Houston Partnership, 2014. Addressing Houston's Middle Skills Jobs Challenge.
2. Ibid.
3. Ibid.
4. Ibid.

training process – this case study will look at how community colleges are achieving this objective.

THE CONTEXT

The First Boom Period

Texas was originally a highly successful exporter of agricultural products with cotton, lumber and cattle ranching the principal means of revenue generation. Houston became a leading commercial centre due to its shipping capabilities, and it was this capability that centred the eventual oil boom on the city. The first significant oil well in Texas was in the town of Oil Springs, with production beginning in 1866. The first economically significant oilfield was the Corsicana field developed in 1894. Global demand for oil as a cheap alternative to traditional lighting fluids drove further exploration across the state of Texas. In 1901, commercial-scale oil discoveries at Spindletop revolutionised the region's economy causing rapid urbanisation and industrial development and attracting new communities to 'boom towns' – towns that sprung up around the oil operations taking place across the state. Spindletop produced an estimated 100,000 barrels of oil a day and in 1902 total annual production was around 17 million barrels.[5]

Throughout the 1910s and 1920s, oil production steadily increased, making Texas the leading oil producer in the United States. Fast growth and industrialisation quickly established Texas as a region with some of the most populated cities in the US.[6] Oil operations and opportunities for work attracted people to areas with high-industrial activity – Houston, as an example, saw population growth of 555% due to oil discoveries between 1890 and 1930 (by 1930 the total population for Houston was estimated to be 292,352).[7] During the oil boom period, Houston became home to a number of oil companies, including Texaco, Humble Oil (which, alongside Standard Oil, would later become Exxon) and Gulf Oil Corporation. In this early period, the oil and gas sector provided economic stability and the revenues from oil extraction and production helped the region avoid the effects of the Great Depression, with Houston being referred to as the 'city that the Depression forgot'.[8]

The state's dominant seaport, the Port of Houston, was seeing large quantities of oil move through it, attracting other industries into Houston and the surrounding areas. Chemical plants, cement plants, automobile manufacturing and steel factories all benefitted from cheap fuel supplies. The growth in industrial activity due to cheap fuel was essential to Houston's ability to withstand the catastrophic effects of the Great Depression.

---

5. Wooster, Robert; Sanders, Christine Moor: Spindletop Oilfield from the Handbook of Texas Online, Texas State Historical Association.
6. 'Population of the 100 Largest Urban Places: 1950'. U.S. Census Bureau and 'Population of the 100 Largest Urban Places: 1940'. U.S. Census Bureau.
7. Texas Almanac: City population history from 1850 to 2000. Texas State Historical Association.
8. Kraemer, Richard H. (2008).Texas Politics (tenth ed.).

The Second Boom and Bust Period

Throughout the 1970s and 1980s a number of political events unravelled, causing booms and slumps in oil prices, and having a significant knock-on effect for Houston's economy. In 1973 the oil crisis, also known as the 'oil shock', saw prices quadruple from US$3 per barrel to $12 per barrel. The reason behind the sharp increase in price was due to Arab oil-producing countries putting an embargo on exports to Western nations in an attempt to influence Western foreign policy. Politically, this move – made by the Organization of Arab Petroleum Exporting Countries (OAPEC) – was framed by the Yom Kippur War and the US stock market crash under the Nixon administration. At the time, the United States was experiencing its oil production peak. This meant that production could only decrease. All these factors impacted negatively on the US economy.

Soon after, in 1979, when Texas experienced another oil boom (due in part to the Iranian revolution causing a sharp decrease in global oil supply) the United States was able to capitalise on the resulting hike in oil prices. Between 1978 and 1980, the price of West Texas Intermediate crude oil increased by 250%.[9] By 1984, a decline in industrial activity and oil consumption had caused a slump – or 'glut' – that brought the price of oil down again. Remembering the 1973 oil crisis and facing another economic slump as production slowed, bumper stickers started to appear with the slogan 'please give us one more oil boom – I promise not to screw it up this time'.

Despite the subsequent 'oil glut' from surplus production – which caused employment rates to drop – the downturn was not as bad as media reports made it seem. Oil prices were still higher than they were prior to the boom and the 'glut' was only temporary. However, these events raised awareness and enlightened leaders to the global role Houston played in energy security. Changes had to be made in how Houston would secure itself against the uncertainties of the global energy market in future.

The Shale Revolution

Another boom began with shale oil and gas recovery. In 2010, it was reported that the development of shale gas resources had generated 600,000 jobs nationwide.[10] According to reports, the energy industry makes up nearly half of Houston's economy despite diversification (this is down from 87% in the mid-1980s).[11] Upstream activity is still important and it is experiencing an increase in activity due to high prices and shale gas opportunities across Texas.[12]

---

9. FDIC: FYI – U.S. Home Prices: Does Bust Always Follow Boom?
10. IHS Global Insight, The Economic and Employment Contributions of Shale Gas in the United States, December 2011.
11. Houston Chronicle, May 3 2013, Houston's Economy – diversified, but still all about energy.
12. http://www.city-data.com/us-cities/The-South/Houston-Economy.html.

Access to large reserves of shale gas that were previously unrecoverable has been made possible by advances in drilling technologies (as well as an elevated oil and gas price) and this is changing the energy outlook of the United States. Shale gas is a hugely important natural resource for the United States and is estimated to make up over 20% of US natural gas production.[13] The Energy Information Administration predicts that by 2035, shale gas will account for 46% of natural gas supplies.[14]

Located in the Texas area, the Eagle Ford shale formation is a major contributor to the Texan economy. Reports indicate that it had a $60 billion impact on the economy in 2012 and supported 116,000 jobs.[15] Major players include ExxonMobil, ConocoPhillips and Cabot Oil & Gas. A study by UTSA (University of Texas and San Antonio) reported that by 2020, 5000 new wells will be constructed, which could potentially create 68,000 full-time jobs. Analysts expect the current amount of unconventional wells to keep rising until 2040.[16] Figures for production show that between January and March 2014, 804,299 barrels of oil and 3496 ft³ of gas were produced each day.[17]

Since the first well was drilled in 2008, the Eagle Ford shale formation has been a source of optimism and job creation in the Texas area. The shale gas boom in the United Sates is driving job growth and there is considerable interest in retraining people, especially military veterans, for new jobs in the growing oil and gas sector. As gas exports rise and domestic generation of electricity continues to lean towards gas and away from coal, the need to exploit unconventional gas reserves will put pressure on the industry to produce trained professionals with the right skills and knowledge to fulfil new workforce requirements.

The strength of Houston's oil and gas sector today is evident in the success is has in producing employment growth and job retention. In the lead up to the global financial crisis, the region's oil and gas sector out-performed nationwide job growth, seeing 6% gains on average.[18] According to recent studies, the oil and gas sector employed almost 300,000 workers in the Houston area in 2013[19] and remains a strong sector in Houston's highly diversified economy, employing twice as many engineers as the national average.[20] Not only did Houston recover much faster than the rest of the United States following the global financial crisis, the recovery was primarily led by the city's energy sector.

---

13. Annual Energy Release Outlook 2012, US Energy Information Administration.
14. Annual Energy Release Outlook 2012, US Energy Information Administration – sourced from Education and Training for the Oil and Gas Industry, The Getenergy Guides, Vol.1.
15. Eaglefordshale.com – http://eaglefordshale.com/.
16. Vidas H, Hugman B. ICF International. Availability, Economics, and Production Potential of North American Unconventional Natural Gas Supplies, ICF International; 2008.
17. StateImpact website – https://stateimpact.npr.org/texas/tag/eagle-ford-shale/.
18. Addressing Houston's Middle Skills Job Challenge, April 2014. (GHP).
19. Ibid.
20. Houston Chronicle, May 3 2013, Houston's Economy – diversified, but still all about energy.

The reason behind Houston's ability to outwit the recession is the fact that it has learnt from the bust of the 1980s. When that happened and many became unemployed, political leaders realised that overreliance on upstream oil and gas extraction did not constitute a reliable economic plan. Since then Houston has become a 'white-collar, energy logistics town', meaning the city is no longer as heavily reliant on its upstream activity.

Whether looking at the past or the present, the region's oil and gas sector has helped to stimulate significant employment and has, to a degree, insulated Houston – and Texas – from wider economic problems. Historically, the employment profile of the region has been inextricably linked to oil and gas. In turn, the industry relies on the community colleges system to train and develop the workforce with the right technical skills for the job. It is for this reason that Houston must continue to build on the successes of its community college system.

THE SOLUTION

ISDs and Community Colleges

Education in Texas is run by independent school districts (ISDs), operating separately from any state or county authority. ISDs have no direct responsibility to any government entity and as such they are largely responsible for their own development. ISDs in Texas originate from the early years of the Texas oil boom. In 1920, the Texas Legislature approved the separation of school from municipal governments as boom towns sprung up around extraction operations. Due to an influx of new residents, the state found it difficult to finance existing schools, so communities took action, deciding to form ISDs that were able to regulate themselves and retain taxes from their specific area, as opposed to sharing taxes across the entire state. This development allowed existing schools to improve facilities. It also gave schools the freedom to source administrative leadership internally from within the district itself, set up independent taxation authorities and access funds without the need to engage with government authorities.[21]

As economic development continued throughout the 1900s, Houston was forced to respond to pressures for more accessible post-school education and for higher and further education to absorb greater numbers of high school graduates. As pressures to absorb more graduates into higher education started to build, high schools had already begun to respond to the needs in their communities by building new facilities and developing 'manual learning' divisions that taught the vocational skills required by burgeoning industries (including the energy sector).

A number of educational developments at this time laid the groundwork for the community colleges system:

---

21. 'The Depot Museum: Oil Boom'. Depot Museum, Henderson, Texas.

- The implementing of universal secondary education
- The rise of teaching as a recognised and respected profession
- The vocational education movement that directly addressed career development

Community colleges emerged from this culture of education reform, mirroring the decentralised education system that characterised school-based education across the country and further empowering local people throughout the United States to undertake some form of post-school education.

The term 'community college' applies to institutions that offer six-month vocational diplomas, one to two year vocational and technical certificates and two year associate degrees. They may be private but the majority are public institutions. The label 'community college' has become an umbrella term for institutions offering the above awards, especially those offering two year associate degrees. However, there have traditionally been a number of differences between various institutions coming under the 'community college' umbrella including two year colleges, junior colleges and technical colleges/institutes.

Much like the rest of the United States, community colleges were formed in Texas to qualify workers for increasingly demanding roles in the industrial sector. A greater proportion of the workforce now needed to be educated in a post-secondary environment and community colleges offered a solution to this challenge. This was especially important in Houston where the discovery of oil and gas had a profound economic, social and educational impact. Like the ISDs, community colleges were organised at a local level as a means of further educating those who were coming into the burgeoning oil and gas industry and as a mechanism to integrate new citizens into that workforce (including the large populations who immigrated to the area in the early 1900s and after World War II).

During the Great Depression period, the emphasis on technical education was integral to the oil and gas industry and, consequently, Houston's ability to ride out the tough financial circumstances that arrested economic growth throughout the rest of the nation. Similarly, the career education initiatives supported by community colleges during the 1970s oil boom and subsequent 'glut' in the 1980s were all integral to ensuring the oil and gas industry did not just continue to take advantage of Houston's upstream sector, but was able to diversify into downstream refining on a much larger scale. However, the needs of the oil and gas industry to do not end with technical competency – they extend into other core subject areas including, but not limited to, language ability. In the 1980s, the wave of immigrants coming into the region could broadly be divided into three groups – those with little technical or language skills, those with low technical skills and good language skills and those with low language skills and high technical skill. English language programmes at community colleges have helped to integrate these immigrant populations into the oil and gas industry through addressing these complex language requirements.

The oil crisis of 1973 and the subsequent decline in price per barrel due to surpluses in 1979, forced officials to reconsider how Houston could withstand global shifts affecting the market. Technical and vocational education was integral to achieving this aim. Through opening educational pathways into the oil and gas industry and through the community college model, not only has Houston been able to diversify economically but the high level of technical training means that the whole state of Texas is able to capitalise on previously unrecoverable shale gas reserves. Developing the technical competencies required for this important unconventional resource is a task that is being met through the community college system.

The standard model of education in Texas remains the ISD, where it is reported that over 1000 public school districts exist. The Houston ISD (HISD) is recognised as the largest public school system in Texas, with 282 member schools and 211,552 students.[22] The decentralisation of the school system in Texas has meant that education policy is dynamic, often employing creative and innovative ways of addressing workforce development challenges. Not only is this approach to education rooted in the oil and gas industry, but its continued value lies in supporting the development of a pragmatic approach to education and training solutions that not only meets the needs of industry, but equips individuals for changes in their chosen career path and supports life-long learning. This approach continues to drive the economic and social development of Houston today. However, challenges concerning vocational education and pathways into oil and gas careers are forcing community colleges to rethink their approach to skills development. This challenge is complex and relates to a number of factors:

- The need to fill a looming gap in technical workers in both the oil and gas sector and related industries that are integral to its expansion
- The need to address the impact of a retiring workforce that threatens to leave the industry bereft of crucial talent and experience
- The requirement to source teachers who have the right balance of industry experience and pedagogical ability and are able to deliver effective vocational education

Today, it is clear that the education and training system in Houston – which has the community colleges at its heart – needs to evolve in order to meet the changing dynamics of the energy industry in the twenty-first century.

The Community College Petrochemical Initiative

One part of the solution to the challenges Houston is currently facing has been the creation of the Community College Petrochemical Initiative (CCPI), an industry-led coalition of community colleges that are working to find new solutions to education and training challenges. One of the founding institutions within this initiative is Lee College. Lee College was established around the ExxonMobil refinery, which has been in operation since 1920. Lee College opened its doors in 1934 with vocational programmes introduced early in 1936. At this time, most enrolled students were employees of Humble Oil (which

---

22. Facts and Figures About HISD – http://www.houstonisd.org/domain/7908.

would later become ExxonMobil). In 1948, Lee College's first advisory council was formed with six Humble Oil representatives serving on that council. ExxonMobil were integral in starting the college's process technology programme, which still retains the name of 'ExxonMobil Process Technology Program', and as part of this partnership ExxonMobil continues to provide subject matter experts to the advisory committees that support technical programmes at the college. The Community College Petrochemical Initiative (CCPI) was started when, in 2013, ExxonMobil engaged Lee College around the need for additional skilled labour in the light of proposed expansions totalling $12 to $14 billion (US). These expansions are estimated to require over 22,000 industrial workers and 1300 to 1500 permanent positions.[23]

Recipients with Steve Pryor: 30 recent recipients of CCPI Petrochemical Scholarships, introduced by ExxonMobil Chemical Co. President Steve Pryor at the August 22 expansion announcement in Baytown, Texas.

Due to their long standing partnership with ExxonMobil, Lee College was asked to lead the Community College Petrochemical Initiative. The initiative involves nine colleges all situated on the Texas Gulf Coast. The workforce development crisis across the industry was such that collaboration between colleges was required (rather than colleges addressing the challenges on their own and at a local level, as is typical within the community college system). The collaboration needed to be one in which the industry and multiple community colleges worked in synergy to tackle the challenge of educating a competent, skilled workforce that could meet current and future requirements.

23. Information provided by Lee College.

New Chevron Phillips employees training at Lee College.

New Chevron Phillips employees training in Lee College's state-of-the-art Pilot Plant.

Initially, it was decided that a steering group should be formed. This steering group would oversee the CCPI and would be led by a representative from ExxonMobil. The first task facing the steering group was to identify barriers to training and developing a technically competent workforce. Out of this activity, a strategy was developed to tackle these obstacles. The strategy defined the following objectives:

- Increase public awareness of the lucrative careers available in the petrochemical industry – part of this was to launch a website to promote available educational pathways to these careers: www.gulfcoastcc.org.

- Increase enrolment on community college programmes that prepare students for roles in the petrochemical industry and other target industries.
- Recruit qualified faculty who have the relevant industry experience to teach technical programmes – this includes sourcing teachers from among professionals nearing retirement from ExxonMobil.
- Create an educational asset inventory for workforce programmes. This will then assist in identifying potential gaps in the supply of education and training provision.[24]

To date, the CCPI network consists of Lee College (the Lead Institution), Alvin Community College, Brazosport College, College of the Mainland, Galveston College, Houston Community College, Lone Star College, San Jacinto College and Wharton Junior College. These colleges work together, through the CCPI, to host or participate in large regional events to raise awareness of educational pathways and recruit students and faculty.

The CCPI is one example of how the community college system stands on the front line of providing technical and vocational training for the energy industry.

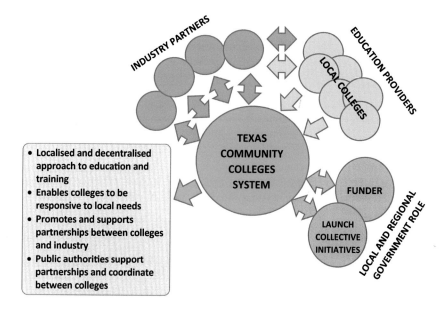

Skilled technicians graduating from a community college have a twofold advantage. Not only do these candidates have the skills needed by the industry within which they are employed. They have also earned credits that can be used towards achieving a university degree, meaning these individuals are able to upgrade their skillsets and increase their knowledge over time. However, the challenge is to increase the number of students enrolled on courses at

---

24. Provided by Lee College, Centre for Workforce and Community Development.

community colleges. State authorities are having some success in this area, but continued efforts to open up more opportunities for higher and further education are integral to fulfilling the needs of industry.

The region's education system is well-equipped to meet the needs of the oil and gas industry and, while challenges remain, the system is built around a number of principals that promote success:

- Decentralisation allows for innovation and faster response to industry needs.
- Partnerships with industry go deep, with examples of companies contributing to the technical aspects of many programmes, as well as having representatives involved with developing curricula for technical and vocational institutions.
- There are significant incentives for industry to collaborate in the development of vocational skills and improve human capacity at the local level.
- Secondary and postsecondary career and technical education receives substantial support from state authorities.
- The emphasis is on supporting core subjects throughout vocational education (e.g. mathematics and English) alongside the development of industry-specific skills.
- The system proactively encourages students to continue education in a postsecondary institution in order to gain the required skills for occupations within the oil and gas and other technical industries.
- The colleges and the industry partners seek to provide access to the technical and vocational education system for Houston's ethnically diverse communities.

Initiatives are expanding, with the aim of attracting new candidates and training a new workforce. The community college system is increasingly viewed as a model for supporting and promoting ex-military/veteran retraining and employment. Some colleges – Lee College is an example – are recognised as 'military friendly' institutes.[25] Community college initiatives to raise the awareness of STEM (Science, Technology, Engineering and Mathematics) subjects are also helping to address the shortage of personnel – a systemic threat to the oil and gas industry.

The Greater Houston Partnership

The Greater Houston Partnership, which has been working to promote the growth of business and economic development for the last 25 years, is made up of some major figures in oil and gas and comprised of over 2000 members. The group seeks to achieve the following:

- Highlighting middle-skills opportunities, increasing enrolment on related programmes and growing the number of potential applicants for those roles
- Preparing applicants for middle-skill positions through training and, in the process, raising employability

---

25. Information supplied by Lee College.

- Providing employers, education institutions and training providers with a mechanism for more effective information sharing
- Creating a data system that can inform workforce development decision-making within a demand-driven environment

GHP operates by bringing together members of Houston's diverse economy and forming committees that directly address the issues and challenges being faced by Houston's industries. GHP have launched a major project, UpSkill Houston, which seeks to improve educational opportunities for Houston residents and develops the competencies required to fill the thousands of occupations that are crucial to Houston's future as the energy capital of the world. The UpSkill Houston action plan ties together various strategies, interwoven to form a comprehensive approach to tackling the issue of middle-skills shortages:

- Sector Councils: the energy sector council will address the critical challenges facing the industry in Houston
- Awareness: GHP will work to change opinions regarding middle-skills professions
- Basic Skills and Employability: the aim is to align training tools and curricula to ensure basic skills are enhanced uniformly, raising the employability of professionals
- Coordination: bring together stakeholders and other parties and communicate information throughout that network
- Supply-Side Synchronisation: working with education and training providers to forge a network where best practices can be shared, information is circulated and industry partnerships can be facilitated.

Historically, the community colleges model has developed as a flexible, dynamic response to the changing nature of the oil and gas industry in Texas. The evolution of this model – and the current approaches being adopted-reflect the ability that this region has for addressing workforce development challenges in the energy sector. As a new revolution (driven by shale) takes hold, Houston's claim to be the energy capital of the world seems safe – and the community colleges can take significant credit for that.

THE IMPACT

The impact of the education and training system detailed above is wide and complex. The maintenance of Texas as a world leader in hydrocarbon extraction, production and processing – and of Houston as its capital – is testament to the ongoing effectiveness of the range of education and training approaches that have been adopted over the years.

More specifically, Lee College's partnership with ExxonMobil represents an apposite example of Houston's success in its ongoing efforts to train a skilled technical workforce. While it must be recognised that many factors have contributed to making Houston a global centre for oil and gas exploration and production, the ExxonMobil/Lee College partnership is a shining example of how fruitful a long-term collaboration can be. Though still in its infancy, CCPI has been successful on a number of levels partly due to Lee College's strong links to the industry:

- More students are accepted onto internships to further practical experience. These internships then lead on to full-time positions with the participating company. The ExxonMobil/Lee College partnership is an example of this.
- Lee College have succeeded in raising awareness of competency-based training (CBT) as an alternative educational pathway.
- The Community College Petrochemical Initiative is helping to develop more flexible pathways into oil and gas careers.
- Lee College is aligning with universities so that technical credit hours can be used towards achieving full degrees. This will increase upskilling opportunities, benefitting the industry as a whole while providing more career opportunities for individuals throughout their employment.
- As CCPI unfolds, Lee College will measure enrolment figures to understand how successful the initiative is, allowing for an evaluation of new students and retention levels.
- More scholarships have been introduced for STEM programmes.
- The college has advisory committees made up of business and industry partners with subject matter experts from ExxonMobil, ensuring on-going communication. In this way graduates can be sure that their skills and knowledge meet the needs of the industry. This model is common to all STEM programmes taught at the College.
- By offering ExxonMobil's retirees 'teaching packages', students can be sure that faculty know the industry intimately.

Apart from the financial support for petroleum-related programmes, the CCPI partnership has been successful because ExxonMobil are active within the college's academic functions, with representatives on college advisory committees that contribute to formulating curriculum.

The GHP, who are using the CCPI as a model for the Upskill Houston project, has been successful on a number of fronts:

- Increasing awareness and raising perceptions of technical career paths and the roles themselves.
- Illuminating the need to fill the widening gap in middle-skill positions across a number of industry professions, especially within the oil and gas sector.
- Improving industry involvement with Houston's social issues with special regard to attracting new members to its workforce, immigration policy and educating a workforce of the future.

THE CHALLENGES

It is difficult to assess the historic challenges relating to the evolution of the community college system in Houston and Texas. Certainly, the boom and bust nature of the oil business has necessitated the education and training community becomes more flexible and responsive.

Today, it is easier to understand what challenges are facing the industry and, in particular, those working to address skills development issues in Houston. There are a number of barriers related to CCPI and Lee College, which include the following:

- Demand for technically competent employees is so high that many students enrolled on oil- and gas-related courses (and other technical areas) are often employed before the end of the programme. Retention of students is difficult but Lee College plan to roll out programmes that provide fast-track training. These programmes will enable students to achieve stackable credentials that can be transferred over into college credit.
- Overcoming perceptions related to competency-based training is another critical obstacle. Candidates generally prefer to enrol on programmes that award degrees and associate degrees, but Lee College and CCPI hope to raise awareness of alternative educational pathways into the petroleum industry and improve the standing of technical and vocational education in the process.
- The need for upgrading the current skills of employees in the sector is essential, especially with mass-retirements on the horizon. Lee College and CCPI plan to make education for existing employees more accessible and available.

Challenges never disappear, even for the world's energy capital, but innovation, commitment and flexibility are all part of building a tripartite approach to oil and gas education, where industry, education and government all form part of an effective solution.

THE COST

To date, the CCPI has not incurred substantial costs for participant organisations. However, ExxonMobil have contributed more to Lee College than they have to any other colleges they are partnered with. The introduction of the CCPI began with a US$500,000 grant from ExxonMobil for the Texas Gulf Coast community colleges to partner together and raise awareness of the need for technically competent workers. After building the initial momentum, ExxonMobil have provided a grant for a further $500,000 so that the programme can flourish.

The Getenergy View

- The story of Houston's development as an energy centre – and, therefore, the story of its struggles to develop a technically competent workforce – runs across more than a century. What we can see in that time is a series of challenges that recur, in line with changes in the industry and in the specific demands the industry places on the workforce. This is a clear illustration of the need for an education and training system to be responsive to change.

- Houston was one of the first places in the world to create and maintain genuine links between the education system and industry. From very early on in Houston's energy story, oil companies recognised the need to play an active role in the education and training of the local workforce. The ability of the education and training system to embrace this type of industry collaboration lies at the heart of the success that Houston has achieved.

- Both historically and today, Houston is an excellent example of how an education and training system can be a mix of the regional and the local. On the one hand, the state and city authorities have played a major role in promoting workforce development through initiatives and policy making. On the other hand, Texas has, over many decades, developed what is in effect a highly decentralised education and training system where local institutions are able to engage at a local level with industry partners in order to meet specific needs.

- Community colleges have played a vital role in meeting the local need for technically competent workers for the energy industry. The community college model is one that is based around a concept of independence and institutional responsiveness. The range of partnerships that colleges are involved in at local level demonstrates not only the willingness of the institutions to reach out and connect with others but also the belief that industry partners have in the ability of these institutions to meet workforce development challenges.

- Although local responsiveness is driven, in part, by the autonomy of the community colleges, there is also clearly a need to coordinate across the education and training system in order to connect what individual institutions are doing to a wider workforce development agenda. This is demonstrated in the initiatives being run at city and state level with the Community Colleges Petrochemical Initiative a case in point.

- Fostering the notion of corporate responsibility in the education and training system has been an important aspect in tying oil companies into playing an active role in Texas. The relationship between state and city authorities, the industry and the community colleges has to be equitable. This means that models of co-funding need to be developed to ensure the responsibility for building a technically competent workforce is spread evenly amongst all partners. Evidence suggests that, over time, these models have been effectively implemented in Texas and that the relationship between the three key protagonists is, by and large, healthy.

A Note on Sustainability

The community college system that lies at the heart of the approach set out here has been running for decades. The long standing nature of the partnerships between colleges and industry are testament to the sustainability of the model. The simple driver behind this success is that the system has consistently produced competent individuals for the oil and gas workforce.

A Note on Replicability

In principal, the system of local college education and training supported by industry is one that is readily replicable in other parts of the world. However, it does rely on there being colleges that are suitably equipped, well-staffed and that have the courses and curricula that match industry needs. As the system has been developing for many years, it may take time for such a model to be successfully replicated elsewhere.

A Note on Impact

If we judge impact against the success and strength of the industry in Houston and Texas, we may wish to rate this as one of the primary examples of effective technical and vocational education in the world. The ability of the system today to respond to the new demands placed on it from the burgeoning shale industry will further demonstrate the impact the system continues to have.

Case Study 5

Kasipkor, Kazakhstan

Creating Vocational Institutions in Kazakhstan for the Development of Human Capacity in Partnership with Leading Education Providers

With thanks to Aizhan Akhmetova, Deputy CEO, Kasipkor

THE MOTIVATION

Kazakhstan is in the process of major reform and modernisation programmes aimed at growing its economy. Kazakhstan's economic strength can be attributed to two factors: first, the rich oil and gas reserves that the country boasts and, second, its strategic position as a supplier to both emerging economies and developed nations. However, Kazakhstan is currently facing a shortage of adequately skilled professionals with the right competencies to support growth in the oil and gas industry – as well as in other sectors – reflecting historic failures of the technical and vocational education system to produce technically competent people. This skills shortage is acutely evident in the oil and gas industry with the sector experiencing increasing demand for workers who are field-ready.

While Kazakhstan enjoys steady economic growth, issues concerning the effectiveness of the vocational education system and the widening gap between labour market supply and demand represent a serious challenge to continued economic success. The government of Kazakhstan has recognised the need to respond to this challenge, both for the burgeoning oil and gas sector and for the wider economy. In 2011, acknowledging the shortcomings of the vocational education system, the government began to implement a number of policy changes designed to promote and support skills development across the country. As a core part of this process, the Kasipkor Holding

Education and Training for the Oil and Gas Industry: Building A Technically Competent Workforce.
http://dx.doi.org/10.1016/B978-0-12-800975-8.00005-8

Company was established and charged with the task of developing and implementing high-quality vocational education and training across all industrial sectors in line with international standards. This case study highlights the work of Kasipkor and specifically focuses on the establishment of the Atyrau Petroleum Educational Centre (APEC), a flagship institution designed to educate and train the next generation of Kazak oil and gas professionals.

THE CONTEXT

Kazakhstan has a relatively small population of just under 18 million, as of 2014, with an average age of 28 years for males and 31 years for females.[1] Kazakhstan has a diverse population, with the dominant groups being Kazakhs (63.1% in 2009) and ethnic Russians (23.7% in 2009),[2] living alongside Ukrainians, Koreans and Germans. National GDP stands at $224.9 billion in 2013.[3] According to recent reports, Kazakhstan is making major strides towards achieving its goal of socio-economic modernisation and this process has been largely funded by revenues from the country's vast oil and gas reserves. Kazakhstan has the largest proven hydrocarbon reserves in the Caspian Sea region amounting to an estimated 30 billion barrels.[4] Oil and gas production contributes to half the country's export revenue and over 30% of total GDP,[5] making it the second largest producer among the post-Soviet republics.

Kazakhstan's largest oil projects are the Tengiz field, on the northeast shores of the Caspian Sea and Karachaganak, located on the Russian border. Together, these reserves account for 40% of the country's oil and gas output.[6] With the discovery of commercial-scale reserves in the Kashagan field amounting to an estimated nine billion barrels, Kashagan is set to become the next major area of Kazakh production.[7] The Kashagan offshore field is known to be the largest oil field outside the Middle East and will make a significant contribution to output rates when production resumes in 2016. The Kurmangazy field, situated in the north, is also an important field with significant recoverable reserves. Additionally, Kazakhstan has access to smaller reserves, some of which are located close to the Chinese border in the east.

This abundance of oil and gas has made Kazakhstan one of the main producers in Central and Eastern Europe and has driven international interest in the region. According to some sources, since its independence from the Soviet

---

1. CIA World Factbook.
2. Official statistics from census reports.
3. Ibid.
4. According to EIA analysis – http://www.eia.gov/countries/cab.cfm?fips=KZ.
5. PwC, 2013. *Helping Energy Companies Succeed.*
6. According to EIA analysis – http://www.eia.gov/countries/cab.cfm?fips=KZ.
7. Dmitry Solovyov, Jul 30, 2014. *Kashagan's oil output restart date shifts deeper into 2016.* Reuters.

Union the region has attracted over US$126 billion in foreign investment.[8] Major international operators include Lukoil (who operate under the subsidiary LukArc), as well as ENI, Total and BG Group. ExxonMobil and Chevron, both shareholders in the Tengizchevroil joint venture – which reported an output of 27.1 million tonnes in 2013[9] – are also active in the region. Kazakhstan has become a focal point for energy security in Europe and Asia, not least due to its strategic position as a supplier to three BRIC nations – China, Russia and India.

A series of modernisation projects have received investment from companies in France, Germany and Italy, all recognising that Kazakhstan is an emerging economy that is attractive to international investors.[10] Policy reforms have also played a crucial role in the development of business, healthcare, legal issues and education. Kazakhstan has exercised careful control over the oil and gas industry through a combination of workforce nationalisation and joint ventures with International Oil Companies, but the continued success and growth of the oil and gas industry is now facing challenges related to the technical competency of the local workforce.

Education is on the front line of Kazakhstan's strategy to energise its economy, especially as it looks towards becoming one of the top 30 developed nations by 2050. Since its independence from the Soviet Union, the government has successfully forged its own approach to economic development and social mobility and the country is now seeking to address issues pertaining to education and training.

The Ministry of Education and Science recognised that the system of vocational training was, in many respects, behind those of developed countries and the implementation of new methods and approaches was imperative to wider economic modernisation efforts. According to the World Bank, an inadequately educated workforce ranked 3rd on the top 10 challenges to business in the region.[11] Kazakhstan has a strong history of educational strength with high rates of enrolment at every level of the system. However, previous approaches to training and developing technical skills reflect an outdated model that predates Kazakhstan's independence.

The key barriers preventing the development of an effective system of technical and vocational training have been identified as follows:

- The need for teachers who possess both the pedagogical skill and technical experience necessary for high-quality training.
- The requirement to develop policy that accurately reflects Kazakhstan's identity as an independent, rapidly emerging economy, especially in the realm of Technical and Vocational Education and Training (TVET).

---

8. PwC, 2012. *Russia & CIS Express.*
9. Reuters, 2014. *Chevron-led Kazakh oil firm hits record output in 2013.*
10. PwC, 2012. *Russia & CIS Express.*
11. http://www.worldbank.org/en/country/kazakhstan/overview.

- Continuing and building on the dialogue between the oil and gas industry and education sector to inform the TVET curriculum and help educators meet the needs of the industry – in the past, tight control of the TVET system led to discrepancies between what was required by employers and what vocational training offered students.
- Sourcing and supplying the necessary equipment that enables institutions to offer the practical experience needed for developing competencies for the oil and gas sector.
- Changing perceptions of vocational education and training and making this an attractive alternative to other educational pathways and careers.
- The need to strengthen employer confidence in students graduating from the TVET system.
- A more strategic approach to the development of vocational centres to increase student access on a national scale.

These challenges are not only important to Kazakhstan in its continued development, but they represent an opportunity for international collaboration in a region that is set to be a leading global oil producer and is emerging as a vital country in terms of energy security.

Now that Kazakhstan has reached a certain level of economic and educational development, the government wishes to build on that success by implementing new approaches that are built around international collaboration. There are three major issues that are driving demand for a renewed approach to TVET training and competency-based education:

- A fast growing economy: with an increase in demand for skilled professionals to work at the heart of Kazakhstan's economy, a major concern for Kazakhstan-based companies is the growing shortage of competent workers.
- Legislation in 2004 to strengthen national control of oil reserves and production capabilities: ensuring the economic benefits of a strong oil and gas sector reach the wider population relies on a strong vocational education and training system that can support the nationalisation of the oil and gas workforce.
- A TVET system unable to meet the needs of the industry: as Kazakhstan forges its identity as an independent country, it is still grappling with the management of its social institutions. Policy that reflects Kazakhstan as an autonomous region is still being developed.

It is against this backdrop that the government of Kazakhstan established Kasipkor and, under that banner, set about the establishment of centres of excellences for the oil and gas industry.

THE SOLUTION

As a response to the shortcomings of vocational training in the region, the government established Kasipkor Holding Company in 2011. This was a radical move to put technical and vocational training into the hands of a new entity that would be charged with achieving the government's goal of creating 'high-level, high-quality and high-prestige technical education meeting international standards'.[12] The idea of looking outward towards international partners who would bring global standards of vocational training was a novel concept within the context of Kazakhstan's centrally controlled vocational education system.

Kasipkor have a 10-year-plan that sets out their aims to tackle skills shortages across all areas of industrial activity. Their approach seeks to meet market demands for better qualified workers who have the right forms of specialisation. To do this, Kasipkor are modernising the structure and content of vocational training – both in regards to the content of the curriculum itself and the quality of vocational teaching. It has been essential to bring the public and private sectors together and Kasipkor benefits from its close ties to government in this regard. The holding company's Board of Directors consists of, among others, the Deputy Prime Minister of the Republic of Kazakhstan, the Minister of Education and Science and the Director of the Department of Technical and Vocational Education.

Kasipkor's flagship education initiative within the oil and gas sector is the Atyrau Petroleum Educational Centre (APEC). This institution was opened in 2013 and is underpinned by a close collaboration with the Canadian Southern Alberta Institute of Technology (SAIT Polytechnic). The centre represents one of the first steps in modernising vocational education and training for the extractive industries.

APEC is one of what will be four regional centres entirely devoted to the training and retraining of oil and gas personnel and has been developed with certain goals in mind:

- Enhance the Atyrou region's oil and gas industry through the education and retraining of specialists.
- Promote the image of the oil and gas sector and vocational education amongst potential candidates by providing considerable financial incentives, free training and free accommodation.
- Utilise considerable public investment into the project in order to purchase state-of-the-art equipment. APEC is housed in a new building in the Atyrou region that has been funded by the Ministry of Education and Science.
- Introduce international standards and competency-based accreditation so that the national workforce is competitive within the global market.
- Provide an attractive and practical pathway to careers in the oil and gas industry.

APEC represents the new approach to vocational education in Kazakhstan that Kasipkor was established to pursue. Not only will it help to develop competent technicians for the oil and gas sector, it will – if successful – improve the profile of vocational education across the country. The involvement of SAIT Polytechnic is essential if this goal is to be achieved. SAIT Polytechnic is a world-class institution and the oldest of its kind in Canada with a long history of supporting skills

---

12. Information reported by the OECD and provided by Kasipkor.

development in the extractive industries. This means that Kasipkor will benefit from the deep well of experience that SAIT can offer as well as learning from the relationships that SAIT has built up over the years with international oil companies including ExxonMobil, Chevron, Shell and others. The instructors at SAIT (who will initially train candidates and mentor local teachers) offer an invaluable source of technical, industrial and pedagogical expertise.

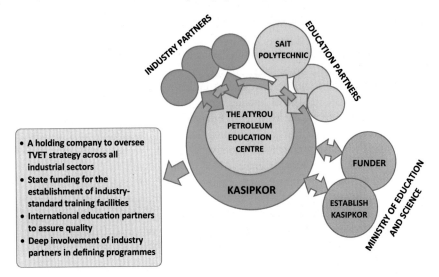

In 2013, the first cohort of students were admitted into APEC, 60% of whom were local to the Atyrau region. Leading up to 2018 Kasipkor aim to build the college to full capacity, taking on 900 students[13] who will be trained in 12 oil and gas specialities identified as being in demand within Kazakhstan's energy sector.

A laboratory in the Atyrau Petroleum Education Centre.

---

13. Information supplied by Kasipkor.

Students enjoy leisure time in the Atyrau Petroleum Education Centre.

The exterior of the Atyrau Petroleum Education Centre

In 2014, APEC implemented a range of measures to attract candidates to study at the institution and, in the process, impact on the negative image of vocational education prevalent in Kazakhstan:

- Students receive full support for accommodation and financial needs, making the institution an attractive option to people in a region where vocational education is not seen as a viable alternative to university study.
- Candidates are selected via an assessment process that aims to gauge English ability alongside proficiency in engineering and science. This process ensures that candidates have the ability to meet the demands of the courses on offer. By establishing high standards from the outset, the institution is working to reverse the mind-set that vocational education is less desirable or less rigorous than other educational options.
- Initially, APEC rolled out courses focused on three specialisations looking at broad processes across the oil and gas industry (there is more detail

about these areas below). All courses have been developed in conjunction with SAIT and with the involvement of industry bodies and associations.

160 students have been enrolled on the Pre-foundation Programme which is designed to ensure candidates can either go on to continue their studies at APEC or can find employment within the industry. The programme was launched in September 2014 in cooperation with the British Council, PetroED and GSE Systems. The British Council are delivering and assessing this programme.

The curriculum model that has been implemented at the Centre reflects the demand-driven nature of training for the oil and gas industry and is based on an analysis of state and corporate training systems. All programmes have been developed in partnership with experts from each individual technical field. While the centre launched with courses in three specialisations, Kasipkor plan to introduce more specialisations to APEC in the lead up to 2017.

The current curriculum covers the following key areas of study:

- *Petroleum Engineering Technology* – this programme covers many aspects of oil and gas operations including exploration and evaluation of oil and gas, drilling and production, well operations (including completion and exploitation of wells), transportation, production processes and refining of products and general industrial control. Students start with a 1 year foundation course followed by a two year petroleum engineering course.
- *Production Field Operations* – this programme involves training for upstream exploration and production. Following the 1 year foundation course, the production field operations course lasts for a further year.
- *Instrument Technician* – this programme enables students to specialise in the maintenance of control instrumentation. Again, it is a one year course, preceded by a 1 year foundation course.

In addition to these three areas, APEC has plans to introduce the following courses by 2017:

- Chemical Engineering Technology
- Power Engineering Technology
- Power and Process Technology
- Gas Engineering Technology
- Drilling Technology
- Instrument Engineering Technologies
- Electrical Engineering Technology
- Industrial Equipment
- Heavy Industrial Equipment

Candidates that graduate from APEC will receive two certifications: a Kazakhstan-recognised national diploma and an internationally recognised certificate from SAIT.[14] Students enrolled on the one year programme will receive SAIT accredited certificates recognising their skills, while students enrolled on the two year programme will receive diplomas upon completion.

All programmes are modular and each module aims to develop a specific set of competencies related to certain types of work. Ensuring that students have opportunities for hands-on experience is at the core of this approach. This is achieved in two ways: first, students have access to modern equipment identical to what is currently used at oil and gas production sites, all of which is integrated into APEC's laboratories. Second, students are required to undergo industrial training at production sites on live equipment. This is undertaken with instruction from experienced supervisors.

The process of equipment selection for APEC was subject to significant research and approval. Firstly, SAIT conducted research into the production cycles of Kazakh companies, considering what equipment was being used in specific areas of production. Kasipkor then developed a list of the required training equipment, which was subsequently approved by industry partners. Kasipkor have arranged the purchase and installation of equipment to support for APEC's Instrument Technician and Petroleum Engineering Technology programmes, including:

- Process Training Workstation
- Calibration, Pressure, Level, Flow, Temperature Devices
- Workshop and Analysis Instrumentation
- Process Control Laboratory
- Chemical Engineering and Water Treatment Laboratory
- Thermodynamics Laboratory
- Electrical Motors and Power Laboratory
- Fluid Mechanics Laboratory
- Thermotronics and Refrigeration Laboratory

Future plans for technology and equipment include the installation of a drilling equipment laboratory, a wellsite equipment laboratory and a petrochemical and fluid process operations laboratory.

Industry involvement is key to the model of training Kasipkor and SAIT have jointly developed. By introducing an Industry Council to oversee operations, APEC has become an open forum for the exchange of ideas between educators and industry. The Industry Council at APEC includes representatives from the oil, gas and extractive industries as members and they contribute to all aspects of training improvement. Everything from the formulation of modules to the purchasing of specialised equipment is reviewed by the Industrial Council. The Associate Dean and academic staff of APEC are also involved in this process with the chairperson voted from among the oil and gas professionals who sit on the Council.

---

14. Ibid.

The educational model works by separating final assessment from training meaning that employers have the final say in awarding the qualification. In this way, training is constantly measured against the needs of employers and this ensures that the industry is at the heart of the training process.

THE IMPACT

The APEC institution admitted the first group of 300 candidates in 2013. A new cohort joined the institution in August 2014. This makes it difficult to gauge the impact the institution is presently having on the industry and it will be some years before graduates from the centre become integrated into the workforce. However, the following indicators are suggestive of the achievements so far:

- The approach taken by Kasipkor and SAIT is one that blends international experience with the particular needs of Kazakhstan's oil and gas industry and this is being hailed in Kazakhstan as a model to be followed within other sectors of Kazakhstan's fast-growing economy.
- Kasipkor plan to build on the short-term success of APEC by establishing 15 additional TVET institutions located across Kazakhstan offering training across the vocational spectrum. The plan involves building a pyramid-like structure of educational institutes consisting of world-class colleges, interregional centres and a number of partner colleges located throughout Kazakhstan. Of these 16, 2 are located in Astana and Almaty and these will offer courses in 6 areas, including energy and engineering.[15]
- APEC is one of the four highly specialised interregional centres adapted to training students in specialisations most in demand locally. Other schools include the Ust-Kamenogorsk and Ekibastuz, both of which field-train candidates in machinery and energy, respectively. Ten partner colleges will provide courses on mechanical engineering, energy, machinery and other oil- and gas-related topics. These partner colleges are responsible for the development and execution of programmes but will receive support from Kasipkor in developing specialised training programmes, the provision of required equipment and training for faculty members.

The Kasipkor and SAIT Polytechnic partnership represents an effective example of how to form the organisational structures and curriculum of a TVET system. SAIT's experience and programmes suit the needs of Kazakhstan's TVET system as it continues to develop and strengthen. Future partnerships beyond APEC, extending to other oil- and gas-related colleges and into other vocational fields, would be advantageous and this seems to be the intention of both parties.

---

15. Ibid.

It is worth noting that the overall project to improve Kazakhstan's TVET system through the implementation of policy changes, the creation of Kasipkor and the development of APEC and other institutions is considered to be successful in terms of what has been achieved so far. In the third quarter of 2014, The World Bank reported that 'the project [of modernising Kazakhstan's TVET system] continued to make considerable progress towards the development objectives with the key outputs for an improved policy framework and enhanced institutional capacity'.[16] Indicators show that many key targets are being met or are progressing well, despite some of the efforts being described as 'ground-breaking'.

Kazakhstan's government has set out initiatives to move the country into the world's top 30 most developed nations by 2050. Improvements to the TVET system and the ability to produce field-ready, competent professionals for the oil and gas sector will be integral to those plans. The continued growth in oil production, the ability to address skills shortages and industry needs and ensuring competency throughout the supply chain will all have a positive impact on economic and industrial development.

APEC is still relatively young, but the success of the institution will depend on several key considerations:

- Ongoing discussion with the business community to isolate and identify needs and competency gaps amongst Kazakhstan's workforce.
- Assessing the future needs of the industry and adapting modules and courses accordingly to ensure that programmes remain relevant.
- Continued collaboration with leading international educational institutions to maintain international accreditation standards. Such partnerships will be necessary to ensure that the quality and efficacy of training and assessment is maintained – this will help to foster positive attitudes towards vocational education.

At the heart of the success of the approach is engagement with industry. While SAIT helps develop the curriculum and programmes, it is the industry experts who map these against their specific needs. Industry participation in the partnership has been managed through the implementation of the Industrial Council (outlined above) and this has provided an effective context for engagement. This will be integral to the continuous improvement and success of APEC's activities in future.

According to the 'State Programme of education development in the Republic of Kazakhstan for 2011–2020', the key quantitative targets for future success are the following:

- Increase the number of employed TVET graduates in the first year after graduation to 80%

---

16. The World Bank, Kazakhstan – Technical & Vocational Education Modernization (TVEM): P102177 - Implementation Status Results Report: Sequence 09 (English), Pp. 2 http://www.worldbank.org/projects/P102177/technical-vocational-education-modernization-tvem?lang=en.

- Increase the total number of teachers who hold the highest categories for national TVET teaching levels to 52% of the total number of teachers
- Ensure 80% of graduates pass employer's independent assessment of qualifications upon first attempt[17]

The future impact of Kasipkor – and of institutions like APEC – will be measured against these targets.

THE CHALLENGES

During the establishment Kasipkor and the launch of the APEC institution, a number of key challenges have emerged:

- Attitudes towards vocational education have been a central challenge for Kasipkor. According to official statistics, only 36% of young people choose to attend vocational colleges, while 63% opt to continue their education in school after ninth grade. Of those who complete their schooling, most prefer to enter university education (37%), 18% decide to leave formal education to go abroad or for other reasons, and 17% enter colleges for vocational education.[18] These figures reveal that vocational education is seen as a secondary option to higher (academic) education and in many cases is not even considered a preferable alternative to leaving formal education altogether. Kasipkor will have to continue promoting the image of vocational education, making it not only a viable alternative economically but also ensuring that the oil and gas industry is attractive to potential candidates. Steps are being taken to address this issue. APEC has the backing of the Ministry of Education and Science and the partnership with SAIT will help to improve the profile of vocational education.
- The standard of English Language ability remains a national challenge that directly impacts on the industry. English is not only necessary for a career in oil and gas, it is essential to students studying at APEC where SAIT trainers from Canada teach in English. In order to help all students access academic materials, Kasipkor established an introductory programme entitled 'An Introduction to the Profession' designed to prepare candidates for entry into the institution, focusing especially on intensive English language training. The introductory programme opens opportunities to socially vulnerable members of society looking to pursue a career in the industry.
- A further challenge for Kasipkor will be the future development of locally based trainers who have the pedagogical ability and technical

---

17. UNEVOC World TVET Database – http://www.unevoc.unesco.org/go.php?q=World+TVET+Database&ct=KAZ.
18. Figures provided by Kasipkor.

skill to educate a competent workforce. During the Soviet period teacher-training departments supported the development of teaching skills and pedagogy. Only a handful of institutions offer this kind of support in modern Kazakhstan.[19] Steps are being taken to address this issue already, with local trainers working as assistants to foreign instructors. It is hoped that this arrangement will allow trainers to adopt best practices and, over time, enable local professionals to replace foreign trainers. SAIT Polytechnic see it as their 'mission' to recruit industry professionals in Kazakhstan to eventually lead the training – these local professionals will shadow SAIT instructors and learn to teach the curriculum. APEC's five-year-plan suggests that foreign instructors, professors and managers will be invited to teach at the centre and that this time period will offer ample opportunity for local experts to gradually replace expat personnel.

- Kasipkor has also identified an important future concern – the retention of graduates within a competitive global market where there is increasing need for trained, competent professionals. The approach adopted at APEC could become a victim of its own success with graduates leaving Kazakhstan for career opportunities overseas.

THE COST

All initial costs for the construction of APEC were met by the Ministry of Education and Science. After completion of the project, management of the facility was handed over to Kasipkor Holding. While partner colleges can expect support from Kasipkor, ultimately they fall under the jurisdiction of local executive authorities. Public spending on new buildings, equipment and financial incentives to enrol candidates has all been critical in the development and early success of APEC. No specific figures are available for the investment in APEC.

Anecdotal evidence suggests that the relationship between Kasipkor and SAIT represents a US$10 million commission for the Canadian polytechnic.

---

19. UNEVOC – http://www.unevoc.unesco.org/go.php?q=World+TVET+Database&ct=KAZ.

The Getenergy View

- Kazakhstan faces significant challenges in nationalising the workforce, not least of which is the demographic challenge of meeting workforce demands with a population of less than 18 million people. The proactive response of the government of Kazakhstan in establishing and funding Kasipkor as a central authority of technical and vocational skills development is an acknowledgement of this challenge and of the need for coordinated action.

- Part of the process has been to look internationally and to create sustainable partnerships with education and training providers to ensure that reform of the TVET system is in line with the standards that industry demands. The recognition from a former Soviet state that the solution to workforce challenges lies, in part, with an opening up of the education and training sector to international competition is an interesting facet of the Kasipkor story.

- The reform process in Kazakhstan is driven at both national and local level. Kasipkor is driving the agenda at national level and is overseeing the long-term strategic plan for the whole TVET reform project. At a more local level, the example of the APEC demonstrates both a sectoral and regional aspect to the solution. It will be interesting to see how this dynamic – between the national agenda and the localised delivery – will play out.

- The language challenges cannot be underestimated, particularly within the energy sector where the majority of companies expect a good level of English and the bulk of technical and vocational training courses are delivered in English. The partnership that APEC has with a Canadian institution furthers the requirement for candidates in the oil and gas sector to reach a requisite standard of English.

- The role of SAIT in the development of the APEC institution has been critical. SAIT have a long history of success in workforce development within the energy sector and they are clearly a valued partner here. However, it is important to note that they have been expected to operate within a Kazakh context and that this has extended to the approach to certification with graduates receiving both a national and international certificate.

- The costs of building the APEC facility have all been met by the government and are likely to be substantial (although figures have not been provided). Anecdotally, the partnership with SAIT has involved an investment of around $10 million USD. This means that, overall, the cost per student will be very high and this calls into question whether the funding model is sustainable in the long term.

- The model of faculty development for APEC is to transition from Canadian trainers to local Kazakh trainers over time. It will be interesting to see how effective this approach is and whether there are incentives offered to SAIT in regard of speeding that transition (bearing in mind that, as a paid provider, a significant part of SAIT's income will be derived from the fees associated with employing Canadian teachers and trainers).

A Note on Sustainability

The commitment of the Kazakhstan government to what is a significant programme of reform is demonstrated by their establishment and funding of Kasipkor. Their plans for the future are ambitious but there is nothing to suggest that continued government funding will not support ongoing improvements to the TVET system for many years to come.

A Note on Replicability

The approach adopted in Kazakhstan is remarkably similar in some ways to what is happening in Saudi Arabia. In both cases, significant government backing has been given to support system reform and the development of industry-specific colleges in oil and gas. Both countries are at the top table in terms of oil and gas revenues – without this level of financial strength, other countries may struggle to replicate the approach.

A Note on Impact

Initial indications are that the reform process is already having an impact in terms of achieving systemic change and establishing the basis for more long-term work-force benefits in future. The real impact will be measurable in the next 5–7 years when graduates from the system have been integrated into the local workforce.

Case Study 6

Caspian Technical Training Centre, Azerbaijan

Training National Technicians for the Energy Industry in Azerbaijan

Chapter Outline

With thanks to Shahin Efendiyev, Compliance Team Leader, TTE-Petrofac

THE MOTIVATION

The expansion of the energy industry in Azerbaijan in the mid-1990s – following the ratification of the 'Contract of the Century' by the Azerbaijani parliament in December 1994 – created opportunities for significant growth within oil and gas exploration and production. As one of the companies active in the country and party to the Contract of the Century, BP planned to expand exploration activities and increase production and needed suitably skilled workers in order to successfully implement these plans.

At the same time, the government of Azerbaijan built mechanisms into the contracts it signed with international oil companies to support the nationalisation of the energy workforce, recognising the need for the energy industry to be a driver for wider economic development and social mobility rather than simply a source of revenue.

As part of the Production Sharing Agreements (PSAs) that each contractor signed with the Azerbaijani government, operators were required to develop and implement plans for the education and training of Azerbaijani nationals so that, over time, the industry would become a significant employer of local talent and would not fall prey to the ex-patriate approach to labour market development that had historically been common in many energy nations, particularly those with under-developed education and training systems.

Education and Training for the Oil and Gas Industry: Building A Technically Competent Workforce.
http://dx.doi.org/10.1016/B978-0-12-800975-8.00006-X

Furthermore, BP – as the operator – recognised that by training and developing nationals – and by doing this in Azerbaijan rather overseas – the company could, over time, reduce their operating costs significantly and simultaneously behave as a responsible corporate entity within the communities where they operated. The challenge was to develop a programme of accelerated development for nationals that could meet immediate and ongoing asset demands.

To achieve success in this programme of nationalisation, BP needed to rethink its strategies for identifying, hiring, training and retaining national technicians. Part of this process involved the development of a robust assessment programme to vet potential candidates and to identify the gaps in their technical competence. To address the issues encountered and to solve both the short-term needs and long-term demands for competent technicians, an effective programme of technical education and training had to be designed and implemented. This programme needed to be housed in a facility that was fit for purpose. This was the genesis for the Caspian Technical Training Centre – the CTTC.

Students undergoing practical training at the Caspian Technical Training Centre.

Students at the Caspian Technical Training Centre.

THE CONTEXT

Azerbaijan has, for many years, been renowned for its hydrocarbon reserves. Lying at the crossroads of Europe and Asia, the country occupies a geographical position that affords it a unique economic and cultural significance. Azerbaijan is a country rich in mineral resources, the most important of which is oil. The main oil fields are on the Absheron Peninsula and the Caspian Shelf. Azerbaijan is the world's oldest oil producing country and has a long history of energy production dating back centuries – it has been described as the birthplace of the oil industry. The country's oil industry experienced a boom at the end of the nineteenth and the beginning of the twentieth century. In 1901, the city of Baku produced more than half of the world's oil.[1] During World War II, the Soviet Republic of Azerbaijan produced approximately 500,000 barrels of oil per day.[2] In the 1950s and 1960s, onshore oil production was augmented by significant offshore exploration resulting in the discovery of several oil and gas fields in the Caspian Sea that subsequently went into production. However, oil production during this period peaked in 1967 and fell away thereafter (although gas production remained steady through the latter part of the twentieth century).

In 1991, Azerbaijan claimed independence from the old Soviet Union. The formation of Azerbaijan's political system was completed in 1995 with the acceptance of the new Constitution of Azerbaijan. The constitution established Azerbaijan as a democratic, legal, secular and unitary republic. After the restoration of its independence in 1991, the country faced numerous and varied problems in its attempts to establish a market economy. The transition to free market principals in parallel with an ailing, centrally planned economic management structure weakened the economy of Azerbaijan, especially during the early years of independence up to 1994. However from 1996 onwards, the economy stabilised and productivity grew as a result of the successful endeavours of the state-owned electric power, fuel, chemistry and oil chemistry enterprises, as well as the emergence of new enterprises in the private sector. This growth in the industrial sector stimulated employment and saw the creation of thousands of new jobs, improving prosperity for many citizens.

Oil provides the primary source of national wealth of the country. The oil industry was, and still is, the leading sector of the country's economy by revenue. That said manufacturing has also seen steady growth over the years since independence with many foreign enterprises and joint ventures now established to manufacture products to international standards and with local companies now supplying a growing domestic market that hitherto relied heavily on imports. This has improved the balance of trade and has had a positive influence on the economy as a whole.

1. Embassy of Azerbaijan – http://www.azembassy.bg/index.php?options=content&id=21.
2. Repubic of Azerbaijan, Ministry of Foreign Affairs – http://www.mfa.gov.az/?options=content&id=9&language=en.

The process of independence during the 1990s opened the gate to foreign investment into the Azerbaijani energy industry and created a platform for new exploration activities as well as generating opportunities for international companies to further explore and produce within known hydrocarbon areas. At this time, the government was the main producer of oil and gas but recognised that there were vast untapped reserves and that national operators (and the national oil company) were not capable of fully capitalising on these reserves. Realising the potential economic benefits of an internationalised energy sector, the Azerbaijani government established a series of contracts as part of an integrated oil strategy. In total, these contracts would see around $60 Billion USD of investment pouring into the energy industry in Azerbaijan over subsequent years.[3]

A single contract covering the Azeri, Chirag and deep-water Gunashli region was signed by President Heydar Aliyev and the participating international companies on September 20, 1994. This became known as the 'Contract of the Century'. The awarding of contracts at this time was coordinated, in part, through the establishment of the Azerbaijan International Operating Company (AIOC), a consortium of the 11 petroleum companies that had commercial energy contracts with the Azerbaijani government. This consortium included BP.

The interests of the Azerbaijani government went well-beyond the granting of licenses. The concept of local content was critical to the strategy developed. This meant that companies operating in the country needed to invest in the development of the Azerbaijan population and, furthermore, that there needed to be commitment to investing in education and training. At this time, there was a lack of local education and training infrastructure, a deficit of good local trainers with the right kind of experience and expertise and a shortage of programmes required to bring candidates and employees up to the requisite international standards in terms of technical ability, competence, safety and management.

As a confident, independent democracy looking to build a sustainable economy, create jobs and develop a more effective and accessible education and training system, the Azerbaijani government recognised the importance of utilising the influx of foreign investment into the energy sector as a lever for workforce development, both within the industry and beyond. Having built this principal into the contracts signed with the partners in the AIOC, it was incumbent on those organisations to deliver on the education and training obligations they had agreed to. For BP and its partners in Azerbaijan, this culminated in the establishment of the CTTC.

---

3. M. Wesley Shoemaker, 2014. *Russia and The Commonwealth of Independent States*, 230.

Staff and students at the Caspian Technical Training Centre.

THE SOLUTION

As BP recognised the responsibilities it had for local workforce development, the decision was taken to invest in the establishment of a dedicated technical training centre that would offer a local solution to workforce development. BP went through a number of key stages to make the centre a reality:

- They began by commissioning a document to guide the process of development. This document – 'Competency Solutions for the Caspian region' – set the direction of travel.
- The next stage involved undertaking a tender process. Partner selection concluded with the awarding of the contract to a joint venture between TTE and Petrofac with the success of the partnership in winning the bid attributed to the positive track record of the parent companies.
- A contract was then signed between BP and TTE-Petrofac that set out the requirement to establish and then manage a new training facility. TTE-Petrofac would also be the principal training provider.
- The centre to house all training activities was then designed and built.
- As this was happening, the content of the programme was developed in line with the requirements of the client (BP and its partners).
- The next step was to recruit appropriate staff to the centre and to streamline the approach to recruitment.
- The CTTC opened in May 2004.

The vision for CTTC was that the centre would become a world-class training facility for the technical and professional development of Azerbaijanis working for BP and its project partners. The centre was built with the capacity to train 400 operations and drilling technicians a year within its main facilities with all trainees trained to international standards. The centre has a number of key features that enable it to achieve this:

- *An experiential learning simulator* installed in a Drilling Control room facility in order to train drillers to safely and efficiently operate the modern drilling equipment that is used on the operating platforms in the Caspian Sea.
- *A unique Operations Training Plant* (OTP), which was designed and manufactured in the UK and then assembled and commissioned at the CTTC, to offer technicians the complete experience of starting up and operating the plant in a safe environment.
- *A purpose built Major Emergency Management Suite* designed to train and assess site managers and other site staff in managing emergencies that might arise on their installations.
- *A fully equipped workshop* offering trainees access to equipment that is extensively used within the oil and gas industry, providing the practical experience that is essential for developing competent individuals for technical work.
- *Approximately 30 classrooms* and offices housing both physical and virtual equipment for production and drilling training.

The Centre focuses on the delivery of three core programmes:

- *A Language, Communications and Behavioural Development Programme* that all technicians complete and that develops reading, writing, listening and speaking skills in the English language and introduces technicians to BP's corporate culture. This programme also helps technicians to develop an understanding of the language of the oil and gas business and to explore aspects of safety.
- *A Technical Foundation Training Programme* that gives trainees learning experiences within the classroom as well as practical and experiential learning opportunities using the Centre's on-site facilities. This programme also involves field visits to operational rigs.
- *A Technical Post-foundation Training Programme* that is designed to offer learners a natural progression from the Foundation programme and give them the opportunity to specialise further and deepen their skills.

The English language component of the training programme is essential to the development of the trainees and consists of the following:

- Fast-track opportunities for high achieving students through the use of baseline assessments
- Two core language elements – generic English and technical English
- A variety of English language topics including Health, Safety and Environment (HSE) behaviour training, BP Culture and Ethics and Presentation Skills
- Ongoing language assessments
- Assessment of the final English level that the student has attained

The Technical Foundation Training Programme has the following characteristics:

- Beginning with a student induction followed by baseline assessments to determine the entry level of Technicians onto the Technical programme and assesses current skills, knowledge and suitability for fast tracking
- A defined curriculum that runs over 18 months and includes 7 months English language training followed by 11 months discipline-specific technical training for Instrumentation, Electrical, Mechanical or 12 months for Production disciplines
- A technical curriculum biased towards 'hands-on' practical skills supplemented by theoretical classroom training (split 60%/40% in favour of practical skills training in most cases)
- Ongoing Assessments on both the theoretical and practical aspects of the training and covering all four disciplines (Electrical, Mechanical, Instrumentation and Production)
- Embedded HSE courses in all four discipline areas
- Students inducted on a quarterly basis
- Programmes accredited by UK accreditation body City and Guilds and by the International Well Control Forum
- A training programme curriculum based on a set of industry standards produced by BP and reviewed annually
- Assessments and assessment processes quality assured using a system of both internal quality assurance and external verification (with awarding bodies including City and Guilds, OPITO and JTL)
- Biweekly review meetings between staff to discuss the progress of students and agree on interventions to improve outcomes
- Final Assessments for all discipline areas
- A graduation ceremony to mark the completion of the programme and celebrate the success of candidates

The Post-foundation Training Programme has the following characteristics:

- Launched in 2010 to expand technician development within the established BP competency management system (CMAS)
- Duration of 3 months for Production, 12 months for Instrumentation, Electrical, Mechanical disciplines
- Pursuing the objective of further reducing the time required to develop fully competent technicians from an average of 5–6 years to 3–4 years
- Facilitated by Coach/Assessors and discipline instructors
- Assessments and coaching delivered at both the CTTC facilities and on-site using live plant and equipment

The centre has developed specific courses within a range of discipline areas. These are divided into a number of categories:

- *Production Courses* that include basic operations training, plant start-up and shutdown, steady-state operation and control, plant upset control and management, unit and plant isolations for maintenance, fault-finding, diagnosis and troubleshooting and mechanical, electrical and instrument planned and corrective maintenance execution

- *Drilling Training* that covers practical well-control drillers level leading to the IWCF certificate, practical well-control supervisors level leading to the IWCF certificate, drilling calculations, drilling fluids, casing design and calculations, introduction to drilling and drilling equipment, safety in drilling, drilling technology, stuck pipe prevention, drilling for nondrilling personnel, basic oil spill recovery awareness, mud school, stripping operations and volumetric control and drill string failure
- *Maintenance and Engineering Courses*, which include a set of modules looking at mechanical maintenance and engineering, modules on instrumentation maintenance procedures and systems and modules on electrical maintenance and engineering
- *Health, Safety and Environment Training*, which includes task risk assessment, hazard id and reporting, slips, trips and falls, basic occupational health, lifting operations awareness, first aid awareness, manual handling, working at heights and fall protection, environmental awareness, chemical awareness, behavioural safety training, safety observation and conversation, hazardous area awareness, corrosion awareness and erosion awareness

Once technicians have completed their chosen pathway of study, they are assessed using the BP Competency Management Assurance System (CMAS) and are then able to start working on one of BP's operational sites. They will

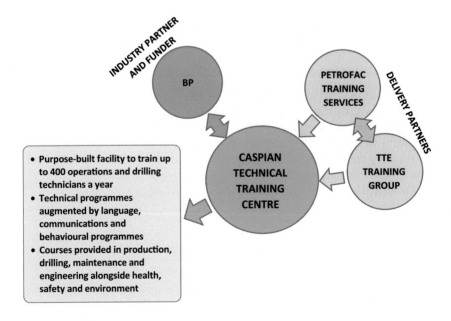

continue to develop their CMAS profile as they progress through their career with the company.

Over the course of the last 10 years, the partners have sought to make incremental improvements to the facility and to the programmes delivered. This process has included improvements to the training approach to meet challenges identified during the early stages of operation, upgrading the equipment at training facility, upgrading the English Training Programme, upgrading the Technical Foundation Programme using feedback from the client and creating a Technical Post-Foundation Training Programme (outlined above).

Furthermore, improvements have been made to the recruitment process (something that had initially proved cumbersome). A complete revamping of the programme was undertaken in order to speed up the process. Today the hiring process on-boards candidates in less than half the time of the original process – going from more than 6 months to 90 days or less. This reduction in time includes better testing, interviewing and assessment of candidates.

A process of feedback originating at the training centre allows the training programme staff to identify knowledge and experience gaps in graduating technicians that can be addressed at the CTTC. Furthermore, site management staff are now more engaged in the progression and promotion of their technicians and whenever deficiencies in the programme content are identified, the feedback comes back to the centre and improvements to the curriculum are made accordingly.

The centre began recruitment of trainers from within the expatriate workforce as there was an acute shortage of locally qualified trainers with the right experience and expertise. However, over time, the plan has been to nationalise the faculty of the centre and to build the capacity of Azerbaijani educators and trainers.

The centre plans to implement a number of courses in future in order to expand education provision and meet the needs of industry. These include:

- Process Safety
- HVAC
- Winter Survival
- OE PA (Operating Essential Performing Authority)
- CompEx Mechanical
- CROER (OPITO)
- Oil Spill Response
- Onshore Emergency Response Training

The centre has also established a limited liability company to offer training services to non-BP customers in the future, demonstrating the rising demand for technically competent graduates across the industry in Azerbaijan (as well as, perhaps, the ongoing paucity of available high-quality technical and vocational education within the industry).

Who is Involved?

BP opened their office in Baku in 1992. Two years later, they became part of the consortium of international oil companies to sign a major contract with the government of Azerbaijan.

The Azeri-Chirag-Deepwater Gunashli field located about 100 km east of Baku is the largest oilfield in the Azerbaijan sector of the Caspian basin. The field is operated by BP on behalf of the AIOC. AIOC formed in February 1995 and currently comprises eight foreign partner companies and includes BP as the largest single shareholder.

BP operates through a number of registered legal entities with offices in Azerbaijan reflecting the evolution of BP's presence in the country and the region since opening their first office in 1992. The principal legal entity is BP Exploration (Caspian Sea) Limited. The company operates under several PSAs and host government agreements signed with the government of Azerbaijan. At the end of 2013, the number of professional staff people permanently employed by BP in Azerbaijan was 3212. Of these, 85% were Azerbaijani citizens.

In order to deliver an education and training solution that was fit for purpose, and that would be economically and socially viable, BP needed partners. To establish and operate the CTTC, they engaged with two companies who continue to be key partners today: Petrofac Training Services (PTS) and TTE.

PTS is a division of energy services company, Petrofac. PTS works with the global oil and gas industry to support the development of a competent workforce. Since 2003, PTS has co-managed the CTTC on behalf of BP and its partners in Azerbaijan. Their involvement has extended to a number of different activities:

- Undertaking an initial assessment of learner styles and levels of technical ability amongst the existing national workforce
- Providing learners and employees with access to production, electrical, instrumentation and mechanical training at foundation, postfoundation and nonfoundation levels
- Developing a tailored oil and gas curriculum that includes English language instruction alongside operations and maintenance training and assessment
- Achieving City and Guilds accreditation for the foundation programme encompassing English language, HSE and discipline-specific training
- Facilitating the design, build and fit-out of the training centre, including the Operations Training Plant that provides hands-on experience of process plant operations in a safe environment

They operate the CTTC as part of a joint venture with TTE International Limited, a UK-based company specialising in the delivery of technical training and consultancy services with a significant footprint in the oil and gas sector. Their joint venture with PTS involves the ongoing management of the CTTC. Petrofac-TTE has a contract to operate CTTC until 2016 and, in 2013, established an 11-month entry programme for BP's annual intake of around

50 Azerbaijani graduate trainees to deepen their technical knowledge, business awareness and communications expertise.

THE IMPACT[4]

The primary objective of the CTTC has, since inception, been to enable BP to nationalise its technician workforce in Azerbaijan. The success of the centre in achieving this can be measured against the following achievements:

- 117 foundation groups have graduated the Centre since 2004
- This means that around 1000 national technicians have graduated from the CTTC's Technician Foundation Programme
- Many other thousands have attended Non-Foundation skills training and competence assessment events with 795 Non-Foundation courses delivered since 2009 and 4817 people certified since 2012
- TTE-Petrofac have made significant progress in nationalising their own workforce in Azerbaijan – of the 104 TTE-Petrofac staff at the Centre around half are now nationals
- For BP in Azerbaijan, the impact has been significant in terms of operation safety with approaching 2.5 million man hours without a Lost Time Injury and nearly four million kilometres safely driven with zero DAFWC (Days Away From Work Case)
- Most critically, 85% of national technicians working on the Azeri-Chirag-Gunashli oilfield and Shah Deniz gas field have graduated from the CTTC, with the Centre providing around 120 new staff for these projects each year
- Informal estimates suggest that through nationalising the workforce in Azerbaijan via this long-term investment in education and training, BP and its partners have, over the course of 10 years, saved around $500 million primarily as a result of the reduction in employment costs associated with expatriate workers

The success of the CTTC approach can also be measured in the replication of the model in other parts of the world by the three major partners:

- BP has attempted benchmark replication of CTTC at three other global sites: the BP Oman Technical Training Centre, BP Iraq (the Rumaila Operating Organisation) and the BP Houma Operations Learning Center in the USA
- TTE-Petrofac are now engaged in developing education and training services to support the Georgia South Caucus Pipeline project. To this end, TTE-Petrofac have opened a branch in Georgia and will be offering the foundation programme to learners at the end of 2014
- PTS is currently working in partnership with Petronas on the INSTEP facility in Malaysia which, like CTTC, aims to offer a world-class technical training curriculum and facilities to support the development of qualified Oil

---

4. All figures in this section supplied by CTTC.

and Gas technicians; they are also working on a joint venture with Takatuf in Oman to establish an industry-leading 'Centre of Excellence' that is slated to become the largest technical training centre in Oman

- TTE is working with Wintershall in Libya to deliver training and assessment of Libyan nationals as part of an on-going professional development programme for petroleum technicians; this will lead to trainees achieving international competency standards

More broadly, BP's ongoing financing and upgrading of the CTTC facility after 10 years of operation is testament to the success of the Centre as a mechanism to develop Azerbaijani nationals into technically competent workers.

THE CHALLENGES

Although the CTTC has clearly achieved a great deal since launch, there have been some challenges along the way:

- *One of the primary issues has been the language barrier*; a principal aim of the programmes that are run at the centre is to give all learners the language skills, behaviours and confidence to start their technical training and, most importantly, to work effectively within today's global environment; all technicians are required to use English during their training as well as during their working lives within BP and therefore there is significant emphasis on developing speaking and listening skills; over the course of establishing the centre, the English language training component has had to be improved in order to speed up the time it takes for learners to reach language competency; furthermore, the centre has implemented more effective language testing of candidates to understand their ability and to fast-track those who arrive at the centre with a good grasp of the language; there are a number of new projects planned to improve the English language level of technicians
- *Finding and recruiting the right candidates has also been challenging*; in part this relates to their age and this can be both positive and negative; if they are young and willing to learn, this is a good thing; younger candidates typically have better language skills, which is also important; however, younger candidates lack experience and this can be a challenge; older candidates bring their experience to the table, which is hugely valuable; however, older individuals are typically less likely to be good learners and also may struggle with the language component; over time, the centre has placed increasing emphasis on effective selection and recruitment in order to ensure that their investment in these individuals is not wasted
- *Identifying and recruiting good quality national trainers has proved to be difficult*; there is a deficit of trainers from Azerbaijan who are able to deliver training in the way that the centre requires (utilising modern teaching methods and practical, experiential approaches) and a question over whether those being recruited are able to adapt to the demands of the centre; over

time, this has become easier although has involved investment in train-the-trainer programmes in order to bring nationals up to speed

The centre continues to explore ways of improving outcomes and has a number of plans in place to do this in future:

- Introduce better quality assurance of English language and technical foundation training programmes
- Introduce more e-learning opportunities into the training programme following the model of starting with e-training, then moving to effective coaching and then to hands-on experience
- Introduce the use of electronic Integrated Safe System of Work (ISSOW) and implement this for maintenance activities at CTTC and for training activities on the OTP

THE COST

Full figures are not available for the ongoing costs of the CTTC although the initial capital investment was estimated to be around $12 million USD.[5]

The estimated lifespan of the centre in 2004 was 5 years and currently a temporary facility is being identified in order to relocate the CTTC facility.

Bearing in mind the estimated cost savings to BP and its partners from the increase in employment of local workers (and the subsequent reduction in reliance on an expensive imported workforce), the costs of the centre are far outweighed by the financial gains meaning that the business model is clearly sustainable.

The Getenergy View

- One of the most striking elements to this case is the estimated cost savings made by the primary investor. This demonstrates that if a locally developed education and training solution is effective in training nationals and those nationals are able to assume positions within the industry, the savings in terms of salaries and recruitment costs are likely to far outweigh the costs of establishing and running the solution itself.
- The CTTC represents a long-term investment on behalf of the industry partners. This proves that, when it comes to education and training, the short-termism that is often a characteristic of investments and decision-making within the oil and gas industry does not work. By investing long term, the solution can become a genuinely valuable asset to a company's activities within a country and not simply a costly exercise in corporate social responsibility.
- The CTTC is an interesting example of an IOC recognising how important it is to seize the initiative and make an investment in the local workforce. Faced with the need to expand operations and employ locals, the response has been fit for purpose and has gone beyond meeting the company's requirement for local content.

Continued

5. Figures supplied by CTTC.

The Getenergy View—cont'd

- There is a sense in which the solution feels like a stand-alone activity that is not connected to the wider education and training system in Azerbaijan. Although it has clearly served the sponsoring companies extremely well (and has supported nationals to become qualified and employed), there is little evidence that the CTTC is collaborating with other institutions locally or is part of a wider programme of Technical and Vocational Education and Training (TVET) reform.
- The challenges that CTTC have experienced are common amongst initiatives of this sort in countries where the education and training system is underdeveloped. Principally, there have been difficulties in identifying good quality trainers to work at the centre. Alongside this is the challenge of finding candidates to undertake courses who are of a high standard in terms of their basic skills. This further proves the point that any solution cannot sit in isolation from the rest of the education and training system.

A Note on Sustainability

The cost benefits to the sponsoring industry partner that have accrued over the course of the lifespan of the CTTC are, alone, enough to suggest that such a solution has significant sustainability. However, the centre was established to meet a particular need within the business. As this need has changed, it will be interesting to see whether funding is continued.

A Note on Replicability

The model of the CTTC is currently being replicated elsewhere in the world (with some of the same partners involved.) This both supports the idea that the initiative has been successful and also suggests that it is readily replicable in other countries. There needs to be demand for local technicians and a company willing to fund the development. Within these parameters, the replicability of the model is high.

A Note on Impact

The impact for the sponsoring organisation has been high with figures suggesting high rates of employment amongst graduates and a surpassing of local content targets. That said, the centre is somewhat isolated from the rest of the education system and so its broader impact has been limited.

Case Study 7

Ogere Training Centre, Nigeria

Developing Nigerian Technicians for the
Agbami FPSO

Chapter Outline

With thanks to Martin Holt, Head of Training, Cegelec Oil & Gas Services

THE MOTIVATION

The establishment of the Ogere Training Centre was driven primarily by the strategic priorities and expansion plans of Chevron in Nigeria. The development of offshore exploration and production activities in Nigeria and, specifically, the arrival of the Agbami Floating Production Storage and Offloading (FPSO) vessel – which was a priority area for Chevron and one that had significant production potential – created a demand for suitably skilled technicians to operate the platform. The company recognised that, over time, the workforce demands were going to be significant.

At the same time, the government of Nigeria was keen to promote the gradual nationalisation of the energy workforce. Historically, the energy industry had provided a significant proportion of the country's GDP but very little of the wealth accruing from the supply chain and from direct employment stayed in Nigeria. With the establishment of a coherent local content policy in the early 2000s, the focus for international companies operating in Nigeria had to shift towards increasing engagement with the local population and greater investments in local capacity building.

As a major international operator in the Nigerian energy industry, Chevron also recognised the need to be innovative in their approach to operating their

Education and Training for the Oil and Gas Industry: Building A Technically Competent Workforce.
http://dx.doi.org/10.1016/B978-0-12-800975-8.00007-1

fields and saw the reputational benefits of investing in the local communities where they operated. Furthermore, it made financial sense to invest in the development of Nigerians and to do this in Nigeria (rather than training Nigerians overseas). The scale of the operation and the long-term nature of their involvement supported their decision to develop a new facility. It was also clear that the local population was hungry for opportunities to work in the industry and that the challenges were therefore to find the right people and educate and develop them in the right way.

THE CONTEXT

Nigeria is the largest economy in Africa and the oil and gas industry is central to this. The industry generates around 14% of GDP and accounts for around 95% of export revenues.[1] Nigeria is the largest oil producer in Africa and has the second largest amount of proven crude oil reserves.[2] In 2012, it was the world's fourth largest exporter of liquefied natural gas and is the largest holder of natural gas proven reserves in Africa and the ninth largest globally.[3] During the early part of the twenty-first century, the United States imported between 9% and 11% of its crude oil from Nigeria. However, this has gradually fallen to around 4% as of 2013.[4]

Demand and consumption of petroleum in Nigeria has grown at an estimated annual rate of 12.8%.[5] However, the domestic market for petroleum products is impacted significantly by the fact that almost all of the oil extracted by international companies is refined overseas with only a limited quantity supplied directly into the Nigerian market.

The industry involves a number of international organisations including Shell, ExxonMobil, Chevron, Total and ENI. The enthusiasm of the international oil companies (IOCs) for participating in onshore and shallow-water oil projects in the Niger Delta region have been dampened by ongoing political instability in the region. This instability has also affected Nigeria's oil production while the natural gas sector has been hampered by the lack of infrastructure to monetise gas (most of which is currently flared).

Crude oil production peaked in Nigeria in 2005 but has decreased steadily since then with security problems caused by the actions of militant groups forcing many companies to withdraw staff and shut down production. Poorly maintained, ageing pipelines and sabotage have caused frequent oil spills leading to

1. EIA, U.S. Energy Information Association – http://www.eia.gov/countries/cab.cfm?fips=ni.
2. PWC, 2013. From Promise to Performance – Africa Oil & Gas review.
3. Ibid.
4. Ibid.
5. NigeriaBusinessInfo.com – Nigerian Crude and Gas Industry.

land, air and water pollution and severely affecting the agricultural livelihood of communities living in nearby villages.

In spite of this challenging operating context, the industry continues to attract international investment. There are several planned oil and gas projects scheduled to begin production within the forthcoming decade although regulatory uncertainty has resulted in the start-up dates for many of the deepwater oil projects being pushed back. This uncertainty has also resulted in a decline in deepwater exploration activity since 2007.

As a result of the significant security challenges facing companies in pursuing onshore and shallow-water exploration and production, offshore oil and gas activities have become an increasing focus for the international energy industry (despite the uncertainties created by the shifting sands of government regulation). It is within this context that Chevron began production activities within the Agbami oilfield.

The $3.5 billion Agbami oilfield project is Nigeria's largest deepwater development.[6] The field lies approximately 220 miles south-east of Lagos and 70 miles offshore, in the central Niger Delta. In 1996, Texaco and Famfa (an independent Nigerian oil company) were granted exploration rights to the 617,000 acre block 216. The discovery well Agbami-1 was completed in January 1999. In January 2000, Texaco drilled the Agbami-2 appraisal well and this confirmed the size of the Agbami structure. Total reserves in the Agbami oilfield are estimated at around 1 billion oil-equivalent barrels.[7] The field came onstream in July 2008 and reached a peak production rate of 250,000 barrels per day in August 2009.[8] A 10-well development programme that began in 2010 will increase crude oil production capacity across the field. Chevron has a 68.15% stake in the Agbami oilfield project and operates the field through its Nigerian affiliate, Star Deep Water Petroleum. The remaining interests are held by Statoil (18.85%) and Petrobras (13%).[9]

The Agbami FPSO was built by South Korea's Daewoo Shipbuilding & Marine Engineering. In 2005, it awarded the contract for engineering design and procurement services for the topsides to KBR and the class contract to ABS. The Agbami FPSO, which arrived at the field at the end of 2007, is one of the largest facilities of this type ever built. It is designed to store 2.2 million barrels of oil and the plan is to have the FPSO on location for more than 20 years. The FPSO is moored in about 4,800 ft of water and at least 40 subsea wells are likely to be required in order to fully exploit the field.[10]

---

6. Offshoretechnology.com – http://www.offshore-technology.com/projects/agbami/.
7. Ibid.
8. Ibid.
9. Ibid.
10. Ibid.

Historically, Nigeria has struggled to ensure that the energy industry employs nationals with the majority of IOCs typically bringing in their own expatriate workers to fulfil technical and managerial positions. In large part, this is a result of a lack of localised education and training infrastructure and a paucity of programmes to support the training and development of individuals to work in the extractive industries. Those nationals who were trained for the industry were typically flown overseas to institutions in the US or Europe. The challenge of developing people in-country was principally one of quality – operational and maintenance technicians need to be trained to acceptable international industry standards and there were no Nigerian institutions able to do this. In spite of the paucity of education and training options, the government of Nigeria recognised the economic and social value of developing a national workforce within the energy industry and set about implementing a local content policy that required international companies to invest in Nigerian talent. It is against this context that the Ogere Training Centre was developed.

THE SOLUTION

Having decided to develop a new training centre to educate nationals for the Agbami FPSO, Chevron set about finding partners who had the right level of expertise to make a success of the endeavour in what was a difficult operating environment. They ran an open tender process that was won by a joint bid from Prime Atlantic Limited – a wholly owned Nigerian Company involved in workforce development in the Nigerian oil and gas industry – and Cegelec Oil and Gas – an oil and gas service company headquartered in France. In order to bid for the Ogere Training Centre contract, they set up a joint venture company, Prime Atlantic Cegelec Nigeria (PACE) in 2005. Having won the contract, PACE set about establishing the training centre.

From the beginning, Chevron had a clear vision for where they wanted to get to but needed partners able to get them there. The training facility would focus on the development of managers, engineers and technicians and would offer them the opportunity to learn in an environment within which they could become competent and qualified to international standards. This would fulfil Chevron's strategic objective of employing Nigerian nationals to work in key positions on the Agbami offshore platform.

In pursuit of establishing the Centre quickly, the project partners took the decision to utilise an existing facility rather than building something new. The site chosen was large enough to offer residential accommodation to students and was far enough away from urban centres to allow the operators to simulate some of the working and living conditions of an offshore oil and gas production facility – in effect, they were able to create a 'closed camp environment' within which students would live and learn.

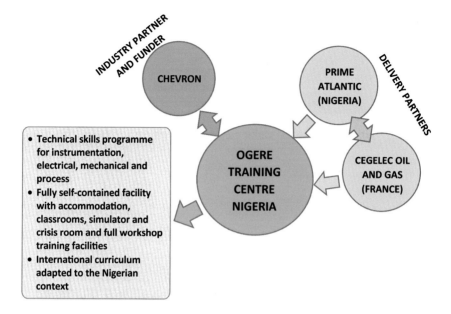

Before starting any training, there was a requirement to implement a competency assurance system so that there was a clear understanding of what the Centre was working towards in terms of competency development. This process involved identifying, for any given role, what competencies are required to perform a task and then building this into the various training programmes offered. This process was undertaken in partnership with the International Human Resources Development Corporation (IHRDC). Having developed a clear picture of the specific industry requirements in Nigeria and mapped this to a competency assurance system, those who were enrolled at the Centre were then measured against the required standards and given training as a result of the gaps identified between their existing competencies and what they needed to fulfil their job.

The first batch of students arrived at the Centre in early 2006. These were predominantly school leavers who had attained a higher or ordinary national diploma. The recruitment process involved putting candidates through a series of tests that posed practical problems and observing how individuals responded to these problems. Those who demonstrated an aptitude for problem solving were selected to go forward.

The programme they undertook had the following features:

- Programme duration was 18 months with students expected to live and learn at the training centre.
- The curriculum was focused on the training and development of oil and gas FPSO operations and maintenance personnel who would be able to assume a variety of positions working offshore for Chevron once they graduated.

- These positions were all at technician level and included process operators and maintenance operators in electrical and mechanical disciplines.
- Part of the programme was to teach students how to read a plan.
- The first batch of students were given the opportunity to experience the practical building of a mini-process demonstration kit so that they were able to develop an understanding of how this equipment was put together.
- This first batch of students benefitted from that experience as it gave them the opportunity to be hands-on and to familiarise themselves with the equipment that they would subsequently train on.
- The Centre adopted an extremely practical approach to the learning experience with the objective of producing competent technical people who were also able to work in teams.
- The aim was that learners would embrace and reflect, throughout the programme, the values that Chevron and the education partners sought to promote – safety and respect.
- Learners were brought on board as a group and encouraged to learn together. The learning experience began with a set of foundation modules that were focused on helping learners understand the industry. Once all candidates had completed these foundation modules, they were then able to decide on the particular discipline they wanted to specialise in.
- Although learners were encouraged to think about which specialisation they wanted to follow, this decision was also informed by the specific needs that Chevron faced. As the programmes were designed to prepare candidates for immediate employment in the field, it was vital that required numbers for each discipline were communicated to the Centre and that uptake of specialised courses reflected this demand.
- By putting all candidates through the same initial foundation modules, the Centre staff were able to assess individuals and identify their strengths and weaknesses. This helped them guide candidates towards specific disciplines.
- New recruits coming into the Centre did not necessarily arrive with a clear idea of the role they wanted to assume within the industry. It was explained to them that part of the process of going through the training programme was to find and follow their own career path.

Today, the programmes delivered at the Centre are focused on a number of vocational training disciplines: process operations, mechanical, electrical, instrumentation and HESS (health, environment, safety and security). All training is accredited by Offshore Petroleum Industry Training Organisation (OPITO). The training model is to offer a range of practical skills alongside the development of knowledge and understanding within a holistic approach that embraces extracurricular activities designed to promote team building. All programmes offered at the Centre seek to:

- develop motor skills and promote mental agility;
- enable trainees to execute tasks safely and consistently;

- promote multidisciplinary awareness; and
- offer assignments based on an ability to perform.

The technical skills element of the core training programme is divided into the following four areas:

Instrumentation

This part of the programme addresses a wide range of topic areas related to instrumentation including:

- instrumentation symbols, signals and drawings;
- measurement basics (including pressure, flow, level, temperature etc.);
- control equipment;
- control systems;
- process control;
- special measurement (ph humidity, gas analysis etc.);
- fire and gas electrostatic discharge;
- working skills (including tubing, soldering and component repair);
- instrument roles; and
- safe systems of work.

Electrical

This part of the programme looks at the basics of electrical training before looking at the following:

- power generation;
- power distribution;
- rotating equipments (including motors, starters and speed drives);
- utilities (including lighting and heating);
- specialist electrical systems (earthing systems, cathodic protection, lightning protection etc.);
- emergency power systems;
- working skills (including measuring devices, testing and troubleshooting);
- instrument roles; and
- safe systems of work.

Mechanical

Alongside an introductory module offering general knowledge on the mechanical aspects of the oil and gas industry, this part of the programme considers the following:

- pipes, valves, gaskets and tanks;
- pumps and compressors;
- reciprocating engines;
- turbines;

- power transmission;
- bearings and seals;
- working skills; and
- mechanical roles.

Process

This part of the programme looks at oil and gas production and processing before covering the following specifics:

- processing systems;
- gas processing;
- oil processing;
- water treatment;
- metering and gauging;
- utilities;
- chemical treatment;
- process operation roles; and
- safe systems of work.

The programme elements are augmented by a dedicated course designed to give students a comprehensive understanding of occupational health, environmental concern, occupational safety and industrial security. The HESS programme involves 90 individual course elements.

The facility itself has been developed to provide trainees with everything they need in order to learn and develop. This includes accommodation in 211 rooms, 22 classrooms, a computer-based training room for up to 24 trainees, a state-of-the-art control room, simulator and crisis room, a fully equipped on-site emergency response health clinic and full workshop training facilities for process, mechanical, electrical and instruments. The replica FPSO that is at the heart of the learning experience is the only one of its kind in Africa and allows trainees to learn how to operate floating production, storage and offloading vessels. The facility also has a swimming pool, gymnasium, tennis courts and a football pitch.

The experience of the learners at the Centre is focused on achieving a number of cross-cutting objectives above and beyond the acquisition of job-specific technical knowledge. These objectives include the following:

- Develop and nurture in every candidate a strong sense of discipline both as a trainee and as employee. This includes a focus on workplace behaviours (like being on time).
- Build a sense of team ethic amongst the trainees and help them to learn to rely on each other and to be respectful of each other.
- Develop a strong sense of personal responsibility amongst all trainees.

- Impress on every trainee the absolute importance of health and safety in everything that they do, both as trainees and when they take up positions within the industry.
- Give trainees an orientation experience that prepares them for the industry and that develops them as if they were already working on a facility.
- Ensure that when they complete their studies and leave the Centre, they are well prepared to start the on-the-job training portion of their learning experience and that they are able to integrate effectively and quickly with the team working on the rig.
- Give every trainee a clear understanding of what to expect from their employment and explain to them the reality of the experience of working on an offshore rig.

The management team created a 'Staircase' model based on the principal that candidates (who, we must remember, are completely new to the energy industry when they first come through the door) are supported to go as high up the staircase as they can. This means that the training programmes delivered at the Centre act not only as a means of building individual capacity but it also function as a recruitment filter that will help the client identify high potentials and fast-track them into more senior roles. The model is implemented using learning milestones with trainees needing to reach an 80% pass rate before moving to the next level. The intention is not to get everyone to the top of the staircase – individuals are supported to find a level commensurate with their ability.

The model followed at the Centre is illustrated below:

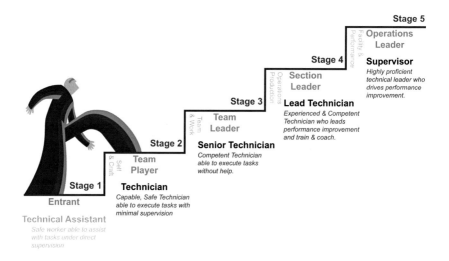

This model also serves as a career planning tool, demonstrating to learners how they can progress over time and giving them an understanding of the pathways available to them for career promotion.

During the establishment and early operation of the Training Centre, there were a number of key success factors that ensured the approach being taken was fit for purpose:

- Learner feedback to management was vital – all learners were given appraisal forms in order to gain an understanding of what they were learning, what they were finding challenging and what they thought could be done better. This was then fed back into a process of continuous improvement within the Centre. The feedback received from the first batch of trainees was the most critical as this was the first time the programmes had been run.
- There was, from the beginning, a major focus on the relevance of the courses to their operations – this was achieved through a close working relationship between with the industry partner and the education and training provider around the development of course content. The communication with Chevron was critical and focused on exploring problems with the curriculum and with the abilities of students and then sorting these out collaboratively.
- The curriculum that was used at the Centre was based on what had been implemented elsewhere. However, the methodology had to be adapted to a Nigerian context – this was the real challenge and also the innovation (as this had not been done before).
- The objectives for trainees when they reached the end of their time at the Centre were that they were ready to move to the next stage of their development – this was the on-the-job training component. On leaving the Centre, trainees are considered capable – after the 12 months on-the-job training, trainees are competent.
- On graduation from the Centre, trainees went into integrated teams on the platform where they spent 12 months learning on the job. This allowed them to demonstrate and reinforce what they had learnt in class. During this process, trainees were evaluated on their performance.

The success of the Centre has been critical as the investment made was high. From the beginning, the approach had to be about genuine impact and this means developing competent technicians who are actually able to operate effectively and safely. It cannot be about giving people certificates. The partnership running the Centre has developed a strong relationship with the client and has been open to discussing challenges and dealing with these as they arise. The long-term vision for the Centre has been informed by the principal that improvements are only made if everyone is open about the problems being faced.

It is also essential that the partnership has taken time to understand the country and the culture of the people, how and why they react in the way they do, how they learn and so on. In this, having a Nigerian company as part of the joint venture has been vital.

THE IMPACT

The primary objective of the Ogere Training Centre was to offer Nigerian nationals the opportunity to train and learn locally to international standards and to support the nationalisation of Chevron's technician workforce. The success of the Centre in achieving this can be measured against the following:

- The centre has seen very low dropout rates since launching in 2006. Those who are selected for training at the Centre are aware of the opportunity this offers them and are very motivated to make it work. Jobs in the industry are highly sought-after and there is no shortage of people wanting to take up positions.
- The first batch of trainees were recruited and trained to work on the Agbami FPSO. Due to the success of the first batch and their evident competency, the client has used the Centre to develop staff for other sites and has split the graduates between different plants across the country.
- In 2009, 3 years after the Centre first started to operate, Chevron invested in a state-of-the-art control room simulator. This demonstrated their commitment to the project and the success they have seen since the Centre established.
- The Centre has trained and certified over 400 Nigerian personnel for the Chevron Agbami and Total Akpo operations in Nigeria to date.[11] It has also provided training opportunities for Shell, Seplat, Midwestern and Chesroc.
- The Centre is making an important contribution to the nationalisation of the energy workforce in Nigeria and has also helped to build local education and training capacity.

As well as having impact on the technical workforce in the Agbami area, the Ogere Training Centre has also had a positive effect on the local community and economy. Ogere is situated relatively far from the larger cities in Nigeria. As a result, the installation of a large training facility in the area has brought employment to local residents – offering significant employment opportunities for support staff in a range of areas including catering, maintenance, cleaning and transportation – and created an atmosphere of collaboration between the Centre and the local village.

In the spirit of commitment to the local community trainees were encouraged to take on projects that were of practical use. One of the major projects

---

11. Figures supplied by Cegelec.

successfully completed by trainees was the repair, installation and maintenance of the generator belonging to the local school. Other similar projects include the installation and repair of water pumps and storage tanks.

THE CHALLENGES

Although the Ogere Training Centre has clearly achieved a great deal since launch, there have been challenges:

- There have been some cultural challenges to overcome in relation to the candidates that have come through the Centre. Many are able, smart people who learn fast and are able to ask the right questions. However, there is often a reticence to admit what they do not know – this fear of being wrong can hinder professional development.
- The Centre delivery partners needed to carefully manage expectations early on. It was vital that there was adequate evaluation in place after the first batch of trainees graduated to ensure there was a clear plan for continuous improvement.
- Dealing with local content legislation can be challenging as the realities on the ground do not always match the intentions. One example of this is that the legislation in Nigeria requires that every expat worker is meant to have an understudy working alongside them. However, it can be difficult to find locals who are willing or able to fulfil these positions.
- Attempts to nationalise the training staff at the Centre have proved challenging. Initially, all the discipline training was delivered by expatriate workers with only one Nigerian Health, Safety and Environment (HSE) trainer. All ancillary and support services jobs were fulfilled by Nigerians so there has been an increase in local employment but it has been difficult to find suitably skilled trainers in Nigeria. That said, this has begun to change with the Centre starting to employ Nigerian nationals as trainers in certain technical disciplines (notably electrical) from 2013. The process is ongoing.

THE COST

Figures for the level of investment going into the establishment and ongoing operation of the Ogere Training Centre have not been provided. The impact of the Centre in Nigeria is partially measurable against the long-term sustainability of the project. The level of industry investment in the Ogere Training Centre is likely to be significant, but when one factors in the contribution to the country's development alongside the cost of expatriate manpower – and the medium to long-term savings that are made as a result of employing a locally trained workforce – the return on this investment is considerable (although difficult to quantify).

The Getenergy View

- This is a highly specialised solution to a very specific challenge. The Ogere Training Centre was established to serve the Agbami FPSO (although has since served other facilities across the country). As such, it is, to an extent, meeting a very particular need. There is a broader question around the extent to which the Centre is making a wider contribution to the talent pool in Nigeria and whether there is any meaningful connection to other institutions.

- The training experience mirrors the working experience. The development of the Centre as a 'closed camp environment' is specifically designed to give candidates a sense of what it might be like to work on an FPSO. This approach ensures that candidates develop familiarity with the demands of the job and acts as a kind of recruitment funnel to ensure that those not willing to live in this way do not take their training any further.

- This is a high-cost model. Although figures have not been provided for the overall cost, the burden of responsibility for funding lies squarely with the main industry client. The solution developed has to account for the fact that the national education system is not producing the people that the industry needs. The Centre has to re-educate candidates and address the failures of the Nigerian education system. This is a major responsibility for an IOC to undertake.

- The notion of competency mapping is key. The Centre was established with a very clear sense of the place that competency mapping plays in the process of training and recruitment. This process encompasses the required competencies for each job role and also the specific competencies that candidates have when they arrive at the Centre. This means that individuals can receive the training they need for roles that they understand. The model also supports recognition of prior learning.

- The importance of culture and behaviour is recognised. A key part of the approach embraces the notion that all employees need to understand and buy into the culture of the company for which they will work. Furthermore, the importance of attitude and behaviour within the workplace is clearly articulated and part of the experience for every trainee.

- The approach requires multiple providers. The Centre is run by a French/Nigerian partnership with qualifications accredited by a British awarding body (OPITO). This demonstrates both the need for companies operating in challenging international environments to spread the net wide in terms of who they work with.

- Career guidance is part of the model. The Centre not only trains individuals, it also helps them understand their strengths and weaknesses and guides them down specific career pathways. The 'staircase model' that the Centre has implemented acts like a recruitment filter that enables people to know what occupation they will do when they leave rather than simply certifying them but leaving them unsure of the role they might fulfil.

A Note on Sustainability

The sustainability of the Centre is based entirely on the sponsoring organisation continuing to fund activities and expansion. Without knowing costs – or understanding cost benefits – it is difficult to assess sustainability. It is also possible that once the Centre has fulfilled its primary role in meeting the workforce demands of the Agbami FPSO, it may not have a long-term future.

A Note on Replicability

The model of the Ogere Training Centre is mirrored by other such initiatives elsewhere in the world. It is essentially a stand-alone training facility that is fully funded by industry for a specific purpose. Its replicability relies on there being a willing industrial partner to both fund and input into the design and delivery of education and training provision. Within this context, the model is highly replicable.

A Note on Impact

The success of the Centre can be measured, in part, by the destinations of graduating students. A majority of those passing through the Centre have found good jobs within the industry with many now working on the Agbami FPSO. Although the numbers are relatively small, it would seem that many of the objectives of the Centre have been achieved.

Case Study 8

Wintershall Libyan Integration and Development Programme

Nationalising the Oil and Gas Workforce in Libya

Chapter Outline

With thanks to Manal Aboujtila, Human Resources Manager, Wintershall Libya

THE MOTIVATION

The German crude oil and natural gas producer Wintershall – part of the BASF Group – is one of the major international energy companies operating in Libya. The company has been active in the country for more than 50 years and has made considerable investments in terms of exploration and production. Wintershall Libya has the licence to operate two concessions both of which are located in the south-east of the country.

Towards the end of the 1990s, the Libyan government began to recognize the importance of promoting and supporting the employment of Libyan nationals across the industry. Although the National Oil Corporation (the state-owned national oil company of Libya) had been established in 1970 and enjoyed considerable influence across the industry, it lacked the capacity to take advantage of Libya's vast hydrocarbon reserves. This led to a number of international oil companies winning licences to operate the fields. As international participation in Libya's energy sector grew, the workforce was predominantly non-Libyan. This was largely due to the fact that the local skills pool was limited and the country lacked the education and training infrastructure to support necessary technical and vocational skills development.

Education and Training for the Oil and Gas Industry: Building A Technically Competent Workforce.
http://dx.doi.org/10.1016/B978-0-12-800975-8.00008-3
115

As an oil company operating at some scale in Libya, Wintershall were increasingly required to hire staff from the local area of Al Wahat, near to the main concessions that it managed. Furthermore, the National Oil Corporation (NOC) decided to play a more active role in the promotion of nationals within the sector and to this end began sending all international oil companies operating across the country an annual intake of Libyan graduates with the expectation that they would be recruited, trained and employed. Initially, this created significant challenges for Wintershall Libya and for other international oil companies who needed to manage the quality and quantity of their intake. The recruitment of Libyans in this way necessitated a well-planned and coherent education and training solution that could bring these candidates up to speed and get them field-ready.

Faced with these challenges – and motivated by a desire to develop a competent national workforce that would sustain the business in the long term – Wintershall Libya decided to establish a programme of training and development in line with the principals of workforce nationalisation that, through active collaboration with the NOC, would put in place a process of selecting and developing recruits in an organised and structured way. This process became known as the Libyan Integration and Development (LID) Programme.

THE CONTEXT

Wintershall in Libya

Wintershall established operations in Libya in 1958 and began producing oil from the Jakhira field in Al Wahat in 1976. During more than half a century of exploration and production in the country, the company has invested more than $2 billion USD and has drilled over 150 wells. Prior to the revolution in February 2011, Wintershall was producing up to 100,000 barrels of oil per day in Libya.[1]

Wintershall operates in the eastern Sirte Basin, around 1000 km south-east of the capital Tripoli and is active in eight onshore oil fields. Their activities in these fields are operated under partnership with the global energy giant Gazprom which has a share of Wintershall Libya's operations. Wintershall also has a share in production from the offshore Al-Jurf platform off the north-west coast, an area it shares with the NOC and the French company Total. In addition, Wintershall has, since 2006, held the licence for another exploration area in the south-east of Libya covering more than 11,000 square kilometres.

Following the revolution of February 2011 Wintershall temporarily suspended oil production due to security concerns although it restarted production activities later that year. Their facilities remained undamaged during the conflict, something the company attributes to their Libyan employees who maintained the production plants. By the end of 2012, production was around 85,000 barrels a day.[2]

As an indication of the company's continued investment in Libya (despite concerns over security in the region), Wintershall commissioned a new 52-km-long

---

1. All figures supplied by Wintershall Libya.
2. Ibid.

oil pipeline at the beginning of 2013 to connect the Wintershall concession C 96 with the Amal field. This represented a partnership with the Arabian Gulf Oil Company (AGOCO) and was planned in cooperation with the Libyan NOC. As a consequence Wintershall became the first international exploration and production company to take over the management of a new pipeline project in Libya.

Post-revolution, the company continued with drilling activities in the As-Sarah field and, as a result, were able to increase production capacity to about 90,000 barrels per day.[3] However, in the summer 2013, production was temporarily suspended upon request of the NOC because the export terminals along the coast were under blockade from local groups seeking to influence the political and economic objectives of the ruling government. Ongoing political uncertainty and instability continue to hamper Wintershall's activities in Libya but their long-term investment in the country and their established technical and human infrastructure suggests that they will continue to play a significant role in Libya's energy future.

The Libyan Energy Sector

Libya's energy sector first emerged in the mid-1950s and, in 1959, the first successful drilling was reported after international concessions were granted in 1956. Soon after, Libya became an oil exporter following the completion of a 167-km pipeline. During the 1960s, Libya's oil production rapidly increased, reaching a peak of more than three million barrels a day in 1969.[4] As a low-income country, the wealth generated by the blossoming energy sector had a dramatic effect. From being one of the poorest nations in the world, Libya suddenly ranked amongst the richest in terms of GDP per capita. However, the wealth that was pouring into the country was not reaching the majority of citizens and this proved a catalyst for the popular support for Colonel Muammar al-Gaddafi's military coup in 1969. In 1970, Colonel Gaddafi effectively nationalised the Libyan energy industry by establishing the National Oil Corporation of Libya.

Over the coming years, Libya's oil production fell steadily from its peak in 1969. By 1973, following Libya's role in the oil embargo to the USA, production had dropped to around half of the three million barrels a day that were produced in 1969.[5] Libya's part in the embargo coupled with a growing opposition to western interests in Islamic states changed world opinion with the west now viewing Libya as a troubled country with an unreliable and unfavourable political leadership. Oil production declined further following the 1982 US embargo against Libya with US airstrikes on Libyan soil in 1986 further damaging international relations. The implication of the Gaddafi regime in the bombing of Pan Am flight 103 in 1987 represented the nadir of Libya's international isolation and further eroded the role of energy production as international markets closed ranks on Libya. Oil production remained at around 1.4 million barrels per day until the turn of the century.[6]

---

3. Ibid.
4. Oil Price, 2011. *Libya's Role in World Oil Production, Past and Future.*
5. Michael Crowder, 1984. *The Cambridge History of Africa, Volume 8.*
6. Petrogas Libya – http://www.petrogaslibya.com/About_libya.aspx.

In 1999, Libya agreed to extradite the named suspects in the Pan Am bombing and in 2003 further reached out to the international community by announcing that it intended to decommission all weapons of mass destruction. In 2004, with UN sanctions finally lifted, international oil companies began to engage more directly with exploration and production activities and oil production began to rise once more. By 2008, production was approaching 1.9 million barrels per day.[7]

Libya's oil production was significantly disrupted during 2011 as a consequence of the civil war and the removal of Colonel Gaddafi from power. As a result, the economy contracted 41.8% in 2011.[8] Although production recovered relatively quickly during 2012, the sector was again disrupted in 2013 by a blockade of ports by groups opposed to the ruling government. Although reserves remain abundant, the ongoing political instability and the challenges this creates for security are likely to hamper oil production and export for some time.

A number of key facts characterise Libya's energy sector:

- Libya is an OPEC member and has Africa's largest proven oil reserves and the ninth largest globally.[9] As such, the country enjoys comparative affluence within the region and is an important contributor to the global supply of crude oil. Prior to the revolution, the Libyan state was generating an estimated 100 million USD a day from oil and gas.[10]
- Alongside proven reserves, Libya is believed to hold large amounts of untapped hydrocarbon resources due to its prevailing geological structures. Much of the country is unexplored so significant investment would be needed to reveal the nature and scale of these unproven resources.
- The NOC – the national oil company of Libya – believes that production of existing resources can be increased by utilising enhanced oil recovery techniques. Their own estimates suggest that additional capacity of around 775,000 barrels per day would be possible from existing oil fields by utilising this approach.
- Libya exports the majority of its crude oil to Europe, with Italy being the principal importer (reflecting historic ties between the two countries dating back to the 1950s).
- In 1971, Libya became only the third country in the world to become an exporter of liquefied natural gas (LNG). The country's gas reserves are relatively small in comparison to crude oil reserves and Libya's one LNG plant was damaged during the 2011 civil war since when LNG exports have been suspended.

Employment, Education and Training in Libya

Within the current political climate, it is difficult to find accurate data in relation to employment. The most recent estimates (estimates that predated the 2011 revolution) suggested that overall unemployment stood at 13% and with the

7. The Oil Drum, 2011. *Libya and World Oil Exports*
8. African Economic Outlook. *Libya 2012.*
9. EIA Libya – http://www.eia.gov/countries/cab.cfm?fips=LY.
10. Youth Policy Fact Sheet Libya – http://www.youthpolicy.org/factsheets/country/libya/.

rate nearer to 50% amongst young people.[11] Historically, the public sector has accounted for a significant proportion of formal employment.[12] A lack of economic empowerment amongst young people is cited by many commentators as being a significant contributory factor underpinning the Arab uprisings across North Africa and this would seem to reflect the situation in Libya.

With around 45% of the Libyan population under the age of 25,[13] reform of education and training at all levels and increasing opportunities for employment have been and will be critical to the success of economic and social policy.

Libya can boast an almost 100% track record in enrolling boys and girls in formal education and helping them complete their 9 years of basic schooling.[14] However, the quality of the learning experience is generally perceived to be low as a consequence of a lack of modernisation over recent years. Despite significant investment in education, school leavers and graduates are not seen as being suitably prepared for the labour market, particularly in the private sector where employers view candidates as lacking relevant competencies and criticise graduates for their negative attitudes towards work. Many employers prefer to hire those with work experience rather than those with advanced degrees, thereby diminishing the value of formal qualifications. The lack of competent Libyans to meet the needs of industry is often cited as the most important obstacle to economic development and prosperity.

Prior to the revolution, around 2.5 million foreign workers were employed in services and other sectors of the economy,[15] despite national youth unemployment rates of between 30% and 50% (depending on which sources one believes). Although specific figures are difficult to identify, anecdotal evidence suggests that those employed in the oil and gas industry are, by and large, non-Libyans, in spite of the nationalisation programme driven forward by the Libyan government.

The labour market in Libya has historically been highly regulated with the public sector as the most important employer and a private sector that struggles within an environment of restrictive labour laws relating to wages and working hours. This has had a negative impact on job creation in the formal private sector and fuelled the continuous growth of the informal sector where labour laws are not adhered to. Furthermore, the last 10–15 years has seen rapid population growth, such that the public sector is no longer able to absorb the active workforce in a way that it once did. The need for a more dynamic private sector where local Libyans have the right skills for the jobs on offer has been evident now for more than a decade. The opportunities offered by the oil and gas industry cannot be overlooked within this context.

---

11. Ibid.
12. Ibid.
13. Using figures from the CIA World Factbook
14. Deutsche Gesellschaft fur Internationale Zusammenarbeit (GIZ) GmbH, 2011. *Libya – Building the Future with Youth: Challenges for Education and Employability.*
15. ETF, July 2013. *Libya: Union for the Mediterranean country fiche.*

Although a vast majority of young people complete school, evidence suggests that the education system has failed to meet the needs of citizens and has not produced enough suitably qualified employees to compete for jobs in the Libyan economy. This has been particularly acute in the industrial sectors (including oil and gas) where the requirement is for competent, well-trained individuals with the right type of hands-on experience relies on a fully functioning technical and vocational education system which can bring educators together with employers. The qualifications acquired rarely enable young people to pursue the careers that they aspire to. Although this is due, in part, to a mismatch between the chosen field of study and available employment, it has been more directly attributed to considerable weaknesses in the quality and relevance of the education and training provided.

It is within this broad context that Wintershall Libya decided to begin a significant programme of investment into building their Libyan workforce. The key factors underpinning this investment were as follows.

- The company was committed to a long-term future in Libya and was determined to build on the success of the previous 40 years of operation.
- The government of Libya was keen to see the energy industry – a sector that had experienced many years of underinvestment and was some distance from running at full production capacity – expand and develop and needed international operators to play a significant role in that expansion.
- The country was experiencing high levels of unemployment, particularly amongst young people whilst, at the same time, employing millions of non-nationals in the industrial and services sectors. The promotion of nationals within the workforce had become a critical policy priority. Within the energy sector, the national oil company was becoming more proactive in taking graduates from the higher education system and placing them with international companies.
- The education and training system failed to adequately prepare individuals for the world of work, with particular weaknesses around technical and vocational education. Greater involvement from industry was essential in promoting improved competency and this applied particularly to the oil and gas industry.

The solution would need to take account of all these critical factors.

THE SOLUTION

In 2000, the Libyan government began to actively implement policies that addressed the integration of national employees into international companies. Prior to this, companies would employ local people at their own discretion and fill positions with non-nationals if they needed to. In light of the paucity of available education and training for the oil and gas sector (and the relatively poor quality of what was available), most technical positions were taken up by expats. As the government began to implement reform, international operators needed to react in a positive and proactive way.

At the request of Wintershall Libya – and in light of the changes that the NOC in Libya were beginning to implement – the Wintershall board of directors decided to address the issue of employing and training Libyan nationals head-on. The first step was to launch the Libyan Integration and Development Task Force. This Task Force was responsible for developing and implementing a structured programme that could increase both the number of Libyans employed and also ensure that those taken on would be able to progress through the company and reach a level of seniority. For the programme to be successful, it needed to involve the provision of industry-standard training that could adequately prepare candidates for working in the field. The LID Task Force set out to achieve these objectives by first gathering all relevant information regarding workforce requirements, the level of candidates and the programmes that would be required. During this process, there was significant engagement with the NOC to ensure that expectations were clear. Once this process was complete, the Task Force presented their recommendations to the senior management at Wintershall.

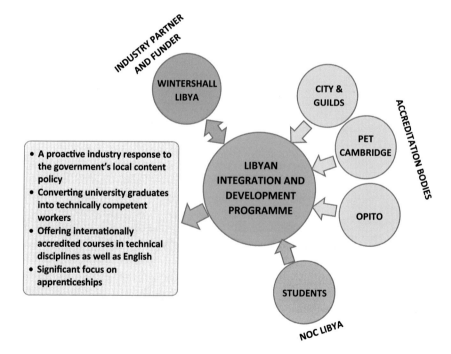

In 2001, Wintershall launched their Libyan Integration and Development (LID) Programme with the simple aim of developing nationals for employment in their operational locations across the country. The programme was based around selecting national employees who had high potential and giving them the opportunity to learn and develop within the company so that they could, in time, replace expatriate employees. A specific aspect of the programme was to take graduates from the higher education system in Libya (who would be sent to the company by the NOC) and train them in technical disciplines as this represented the area of most demand within the business. The company also believed that all employees – regardless of the job they would eventually do – needed to understand the technical aspects of the business before they could progress.

The LID Programme was fully supported by the Wintershall Board and quickly became well-established within the company's operations in Libya. A detailed strategy was developed to guide the progression of the programme and this contained the objective of achieving a ratio of 75% Libyan employees by 2006. The concept was not simply to inflate the size of the organisation in order to employ a quota of nationals. Nor was it to provide low-paid jobs to Libyans as a way of meeting local content requirements. The objective was to provide meaningful training, jobs and qualifications for all employees.

The strategy developed to guide the LID Programme set out the following explicit aims:

- Increase the number, level and quality of Libyan staff employed by Wintershall Libya.
- Increase the proportion and seniority of Libyans working for Wintershall Libya.
- Ensure Libyan staff are able to achieve at least the same standard of work competence as the expatriate he/she replaces.
- Recruit Libyan staff from the Al Wahat area wherever possible.
- Reach the target ratio for Libyan employees of 75%, in line with government requirements on local content.

It is important to note that this programme was established at a time when Wintershall Libya was expanding its operations in the country and was looking to increase recruitment across the board. Within this context, it made sense for the company to look at ways of increasing local participation in order to meet the increased demand for skills. Initially, the process involved employing experienced expatriates within the technical functions of the business and then, over time, hiring locals who would train and learn alongside them. The programme was primarily concerned with the hands-on technical training that was required to fulfil technician positions in the field. However, the programme also looked at other areas of the business – finance, administration, IT – with Libyans trained up to assume these positions within the organisation. Once this had happened, the Programme managers began to look at the technical functions of the business.

The programme was developed over a period of months and followed a series of key phases to ensure that objectives were being met:

- Phase 1: Information Gathering
  This involved researching and analysing job descriptions for those running the programme, developing an organogram, researching and producing manpower projections for the company over coming years and identifying and analysing recruitment plans.

- Phase 2: Programme Planning
  This phase focused on developing the specific training modules and materials that would be needed to support the effective implementation of the programme. This phase also involved the preparation of Personal Development Plans that could be utilised with all candidates.
- Phase 3: Training Implementation
  The third phase took the work completed in Phase 2 and implemented this into the training and human resource operations within the company.
- Phase 4: Improve Job Competence
  The final phase focused on the development of competency standards, designing the support for on-the-job training and then implementing the on-the-job, competence-based training and assessment.

The relationship with the NOC was critical from the beginning. The NOC acts as the umbrella under which all companies in the sector operate. Within the context of the LID Programme, the NOC provided initial training to candidates and would then place these graduates with international companies for further training with every company compelled to accept a set number of graduates every year. On average, the NOC would send 12 graduates per year to Wintershall Libya.

The Training & Development section in Wintershall was given the responsibility for running the LID Programme from its launch in 2001. The training programmes that were run as part of the programme are defined as follows:

Programme	Duration	Qualification
Operations Trainee Programme		
English	1 year	PET Cambridge
OPITO technical training	1 year	City & Guilds level III
Rotation skills	1 year	2 months rotation in each selected discipline
Apprentice Trainee Programme		
English	1 year	PET Cambridge
VQ level II	6 months	NVQ level II (abroad)
Generic training (electrical)	6 months	(On-site)
Generic training (mechanical)	6 months	(On-site)
G&G Programme (Geologists & Geophysicists Trainees)		
English	1 year	PET Cambridge
IOC & Nautilus programmes	1 year	
On-the-job training	1 year	Trained with one of our experienced staff and mentored by a professional geologist

The initial stage of each programme involves assessing the level of English that graduates have in order to gauge their readiness for the training modules (which are taught in English). Those who are not up to the requisite standard receive an intensive English course for the first year of study. This course is based around the Preliminary English Test (PET) accredited by Cambridge English. Candidates can then choose from two specific pathways – the petroleum engineering course and the instrumentation technician course. All courses are a blend of classroom training, technical training and on-the-job training with candidates expected to complete 3 years of study (including their English language training).

Within the apprenticeship training group – which is designed to develop electrical and mechanical technicians – candidates also receive a 1 year English language programme followed by 6 months training in a Wintershall facility to develop their basic technical skills. They are then sent to the UK for a further 6 months in order to undertake a National Vocational Qualification (NVQ) Level 2. Their third year is rotational year – they work across different disciplines under the supervision of senior people within the business in order to gain a wide breadth of experience. This also allows these candidates to see which areas of the business interest them most and to match their competency to a particular job function.

The selection process for candidates is critical as the company invests significantly in each candidate. This is coordinated with the NOC (who are responsible for sending graduate candidates to the programme). These candidates are interviewed by two company panels – a technical panel and a recruitment panel – both of which are looking to identify people who could have a long-term future with the company. When the programme launched, all candidates undertook a technical assessment test. However, the process now relies on face to face interviews with technical panel charged with the job of assessing the technical knowledge and competency of each candidate. Attitude and behaviour is one of the most important aspects of the selection process as this provides the best indicator of whether the company should be investing in an individual.

The NVQ (mapped to the UK National Qualifications Framework) was selected by Wintershall as the international standard for on-the-job training and this is undertaken by all professional field staff. The NVQ standard is evidence-based, well-structured and gives fair and accurate assessment of the competency of the employee. Also, it is an industry-standard qualification that is used by a number of other major oil companies internationally.

Wintershall successfully rolled out the NVQ system in every available discipline in the programme: Electrical, Instrumentation, Production, Well Services, Mechanical, and Laboratory. Wintershall trainers in the field (Production, Electrical, Mechanical) are certified NVQ trainers and assessors.

The Occupational Training Centre (OTC) in Jakhira was completed in October 2007, with five training rooms fully equipped to serve different training groups. All teaching staff working at the OTC are certified trainers and assessors and have qualifications from internationally recognised British awarding bodies. The OTC directly employs three English language trainers, two production trainers, one electrical trainer, one mechanical trainer and an IT trainer.

The OTC was established to enable Wintershall Libya to provide in-house vocational training for their trainees and it has become a critical success factor in the implementation of the LID Programme.

The success of the LID Programme, since launch, is attributed in part to the commitment and support of the Wintershall management team and board of directors. This commitment has been underpinned by a drive to genuinely develop Libyan talent in order to support the future success of the business rather than simply meeting government targets for workforce nationalisation. The project has made good progress towards reaching its goals and all managers and departments within the company have demonstrated commitment to the programme. A further success factor has been the support of, and close collaboration, with the NOC.

THE IMPACT

The LID Programme has met many of its targets and achieved success in increasing the quality, quantity and seniority of Libyan people within Wintershall Libya through recruiting and developing locals into a number of positions that were previously occupied by expatriate staff. Impact can be measured in the following ways:

- The company has successfully reached the initial target ratio of 75% nationals within the overall staffing of the company. Today administrative and financial functions are run 100% by Libyan nationals whilst technical functions are gradually moving towards that target.[16]
- The company now has in place a team of Libyan nationals that was developed through the programme and that includes all disciplines within each of the technical fields.
- Furthermore, some of the graduates from the programme have now reached management positions, thereby achieving one of the key targets set within the initial strategy document.
- Some graduates from the programme have now left Libya and gone on to work in other parts of the business internationally, further indicating the effectiveness of the training.
- Retention rates are also very high within the business with Wintershall being one of the market leaders in Libya in terms of workforce retention.
- The programme itself is now being largely run by Libyans (having started with a predominantly expatriate staff). This has meant that the programme can continue to run during times of increased political tension and instability.

The company continues to invest in the programme, further demonstrating its success within the business, and are now looking at the possibility of developing competency profiles for all roles within the organisation and implementing these into the programme. This will ensure that newly trained recruits will be measured against international standards and receive the recognition of competency that they deserve.

---

16. Figures supplied by Wintershall Libya.

THE CHALLENGES

Wintershall's nationalisation programme has achieved a great deal since launch. However, the initiative has faced some challenges:

- Libya has, over recent years, experienced widespread civil and political unrest creating a hugely challenging operating environment for any companies working in the region. Although the establishment of LID Programme pre-dated these events, the continued success of the Programme today may be undermined by these factors.
- The level of English Language of those candidates coming into the programme has typically been low. This necessitated the 1-year intensive English language training programme for a majority of candidates.
- Many candidates also lacked the requisite soft skills as they were recruited from an academic background where employability was not a priority.
- The lack of well-established vocational training centres in Libya with the facilities to support the programme created challenges in the early stages. Since the construction of the dedicated OTC in Jakhira in 2007, this challenge has been largely addressed.
- The real need, as Wintershall launched their programme of workforce nationalisation, was for skilled and competent technicians. The candidates being sent by the NOC were university graduates who lacked the right type of skills and also had ambitions to be in managerial roles rather than technical positions. Without a positive culture of technical and vocational training in Libya, those graduates needed to be convinced of the value of their technical education.
- Furthermore, the company believes that everyone needs to start their career by gaining some hands-on experience. This gives them grounding in the business but seemed to some graduates like going backwards.
- The high unemployment rate in Libya – particularly amongst young people – resulted in increasing pressure from the government to hire more and more people. At times it was difficult for Wintershall to control the number or the quality of new intakes.
- More recently, the Programme has sought to get the OTC accredited to international standards so that Wintershall are able to deliver and certify other internationally recognised exams within the institution. However, the ongoing volatility of the security situation has hampered these efforts.

THE COST

The overall cost of the programme has been difficult to quantify. It is important to note that for Wintershall, the cost of establishing and then maintaining the programme is absorbed primarily on the basis that the company believes that investing in Libyans is part of their responsibility. That said, the level of investment is likely to be offset by the reduction in recruitment and employment costs.

Expatriate labour is significantly more expensive and the nationalisation of the Wintershall Libya workforce has undoubtedly reduced these costs significantly.

The Getenergy View

- The post-revolution landscape is now very different and it is difficult to judge the impact that this will have on Wintershall's workforce nationalisation programme. However, the need for better opportunities for nationals within key industrial sectors remains. Finding jobs for the people who fought and won the revolution must now be a priority and this will be important for the legitimacy of the post-revolution government.
- The approach taken by Wintershall in Libya can be seen as a model of how a company can respond positively and proactively to government interventions in the industry. The imposition of a local content policy that directly placed university graduates with international oil companies meant that these companies were forced into a position of recruiting individuals with little or no experience of the industry. Wintershall's response to this has been to embrace these individuals, invest in them and give them the opportunity to become valued employees within the company.
- This is a long-term investment. Not only has Wintershall operated in Libya for some years, its approach to workforce development in Libya demonstrates an ongoing commitment to the industry and to the people of Libya that should be applauded. This kind of long-term thinking is rare within the oil and gas sector.
- The commitment of Wintershall has already been repaid. Although the level of investment in the LID Programme is likely to be high, the benefits to the company have been impressive. The number of Libyans now working across the company is high (thereby significantly reducing the cost of employing non-nationals) and the loyalty of this Libyan staff has already been demonstrated in the reaction many showed during the political unrest following the revolution of 2011.
- The starting point for Wintershall was to develop a deep understanding of workforce requirements. The attention paid in the early stages to the specific industry requirements across the business ensured that the programmes developed met the needs of the company and led to graduates from the LID Programme finding permanent positions.
- The relationship with the NOC (the national oil company of Libya) has been critical to the ongoing success of the Programme. By working in partnership with the NOC – rather than grudgingly accommodating their requests – the company has managed to build a solution that meets both their and the Libyan government's objectives for Libyan nationals.
- Converting university graduates has not been easy. The intake of candidates has been made up largely of graduates from academic study. Taking these graduates and giving them what is, in effect, a vocational education has been challenging, particularly in light of the negative connotations that 'vocational' has in this part of the world. The Programme has had to fight hard for recognition and respect within this context.

Continued

The Getenergy View—cont'd

- The Programme has affected all aspects of the business. One of the major successes for Wintershall has been the way in which graduates from the Programme are now working across all parts of the business (with some progressing to senior management positions). This is refreshing in light of the way IOCs often meet requirements under local content policy by recruiting nationals into manual positions in the company simply as a way of meeting quotas.
- The recruitment process has been key. The LID Programme implemented a rigorous process for identifying those candidates that would be worthy of Wintershall's investment. It is particularly interesting that this process no longer involves any kind of written test and is based entirely on interviews. Also of interest is the increasing focus on attitude and behaviour (as opposed to aptitude and competency).
- Wintershall have had to do it on their own. A key challenge for the company in implementing the LID Programme has been the fact that there was little education and training infrastructure to draw on when they started. Ultimately, this has led to the company building their own training facility and delivering all training in-house. This demonstrates the type of issues that face IOCs operating within countries where the landscape and market for education and training is underdeveloped.

A Note on Sustainability

The sponsoring company describes the programme as being part of their responsibility to the local population and community. They have not shared any quantifiable results in terms of the economic benefits of successfully training (and then employing) Libyan nationals. However, the level of commitment to the programme within Wintershall is evidently high. As long as this remains, the Programme will remain sustainable.

A Note on Replicability

The approach taken by Wintershall is instructive in two ways. First, it shows what can be achieved when an international company responds positively to local content requirements and embraces the relationship with the NOC. Second, it demonstrates how commitment from across the business can create genuine impact. In terms of approach, this should be celebrated and, wherever possible, replicated.

A Note on Impact

The impact for Wintershall has been significant – the number of Libyans employed in the business has surpassed the targets set by the Libyan government. Moreover, those graduating from the Programme have gone on to work in senior positions within Libya and in the business overseas.

Case Study 9

The Well Control Institute, USA

Promoting Global Standards in Well-Control Training

Chapter Outline

With thanks to Cason Swindle, Executive Director, WCI

This case was significantly influenced by the Well Control Forum held at the Getenergy Global event in June 2014. Thanks must go to all who took part in that forum.

THE MOTIVATION

The Deepwater Horizon oil spill in the Gulf of Mexico – where an estimated 4.9 million barrels of oil discharged into the ocean[1] – was described by US President Barack Obama as '…the worst environmental disaster America has ever faced'. The incident was reported in the mainstream media as being the biggest unintentional offshore oil spill in the history of the petroleum industry. The incident killed 11 workers and sent 62,000 barrels of oil per day gushing into the sea.[2]

In the aftermath of the Macondo disaster, enquiries established that well control was at fault with evidence that those responsible for the well were unable to respond effectively to the incident resulting in the subsequent loss

---

1. On Scene Coordinator Report on Deepwater Horizon Oil Spill (Report). September 2011.
2. Henry, Ray (15 June 2010). 'Scientists up estimate of leaking Gulf oil'. MSNBC. Associated Press.

Education and Training for the Oil and Gas Industry: Building A Technically Competent Workforce.
http://dx.doi.org/10.1016/B978-0-12-800975-8.00009-5

of life and unprecedented environmental damage. After the failure to maintain well control on the Macondo well, the industry needed to proactively respond to the causes of the incident. The International Association of Drilling Contractors (IADC) and the International Well Control Forum (IWCF), owners of the two well-control training standards, recognised the need to improve the quality of training across the industry and to better match their standards to industry needs.

Within this context, the IADC believed that a new organisation was needed to drive forward the necessary change within the industry. The Well Control Institute (WCI) was subsequently established as an industry body designed to effectively address well-control improvement issues worldwide. Their aim was to create an organisation that could help to define, govern and administer a single well-control training standard (which would be owned by the IADC). The vision that developed for WCI was for an industry-led body to objectively review well-control improvement initiatives around the world and provide leadership and guidance to the industry in order to improve well-control safety and promote better working practices. Part of the approach was to ensure that the governance of this new organisation included representatives from all parts of the industry who had a role or interest in well control and that this should include operators, contractors, equipment manufacturers and well service providers.

As the organisation was established, and discussions began to take place around the role and purpose of the Institute, it became clear that a much wider remit for well-control improvement was possible given the governance structure put in place. The establishment of the WCI is an attempt to address the endemic issues facing the industry in regard of well-control safety and to create an organisation that can act as a promoter of a universally understood set of standards and processes that will ensure competency amongst well-control employees globally and that will, it is hoped, consign incidents like Macondo to history.

THE CONTEXT

The drilling industry is currently in the midst of significant change. This change is underpinned by a number of key factors. First, the industry is witnessing a gradual demographic shift with experienced employees retiring and a younger staff coming on board (although this is a phenomenon primarily affecting the US and Europe). Alongside this, the industry is also witnessing a rapid expansion of the fleet which is placing new demands on recruitment. As well as this significant demand for new hires, the industry also now needs people to reach competency and become field-ready faster than ever before.

Oil rig workers lifting a drill pipe.

An oil rig in the Gulf of Mexico.

Technology is also driving change, both within the industry and within education and training. Job roles require an ever-more diverse set of competencies in line with new methods of exploration and extraction and there is an increasing reliance on simulators as a mechanism to give trainees hands-on experience as they prepare to join the workforce.

Operators and contractors continue to operate within a post-Macondo era whereby regulators are demanding improvements in process safety across the industry. This is being driven, in particular, from the USA where political pressure on the industry to become safer and to demonstrate and assure the competency of staff has continued unabated in the aftermath of the Gulf of Mexico spill. It is also now recognised that the Macondo incident was preceded by other near-miss disasters that should have acted as a warning sign to the industry but that were, in effect, ignored – this has added further impetus to those calling for a radical shake-up of the way that well-control competencies are developed and maintained. Over the course of recent years, the public has become increasingly wary of the industry and the impact of well-control events is highly public. This heightened level of public awareness has been largely damaging to the image of the industry and operators and contractors are now concerned that another event of the size and impact of Macondo will provoke regulators to exert much greater influence across the industry.

There needs to be a step change in safety standards and the regulators are watching. As the 2010 Macondo incident begins to fade into history, governments, regulators and the public are asking what the industry has done to address the failures that led to such a damaging disaster. This pressure is driving change. A number of key organisations – including International Association of Oil and Gas Producers (OGP) and the IADC – have been active in setting out the deliberations of a large number of industry representatives looking at the challenges post-Macondo and this has signalled the direction of travel. These deliberations have focused squarely on the role of education and training and specifically on qualifications, the accreditation of providers and the development of global standards.

Operators and contractors are beginning to recognise and acknowledge the correlation between competency and incidents/accidents and this will be a key driver for the adoption of global standards and curricula across the industry. That said, there is a shelf-life to an event like Macondo – people forget, believe that things have changed and then problems can re-emerge. Within this context, the response needs to be permanent and far-reaching. There is recognition that well-control training needs to improve but there are no indications that companies are willing to increase their level of investment in such training. This means that, in education and training terms, the requirement is to do more with less.

Within this context, there are a number of critical challenges that face the industry:

- There is a demand for new models of competency development that are both effective and timely. Part of this is the concept of continuous learning – training should not be 'an event' that happens every two years and you cannot wait 2 years to learn the lessons.
- On-the-job training (OJT) – which is recognised as a critical element of any effective approach to training in the industry – has to be supplemented with structured standardised programmes in order to move people up faster and with greater competency. That said, there are often issues with access to OJT and a lack of experience within companies to deliver effective OJT. This will hinder competency development.
- There is growing recognition of the importance of human factors in well-control training with around 50–60% of well-control incidents attributed to human factors.[3] However, human factors training is still something that is in the early stages of development within the oil and gas industry and the main focus is

---

3. Figures supplied by WCI.

still on technical training. Technical skills need to be built in alongside human and organisational factors and this needs to be integral to every programme.

- The list of human factor elements that impact on the performance of well-control employees is long and the industry typically shies away from this challenge as it is considered too complex. There are still questions over how human factors training can be effectively integrated into well-control training.
- There are further challenges around knowledge retention with the typical gaps between training experiences sitting anywhere between 2 and 5 years – this is not good enough within an industry that is constantly changing and within a context of steady technological evolution. Many organisations across the industry are failing to adequately implement a culture of continuous training.
- Well control is a diverse business – it involves different processes that need different types of training (ultra-deep water vs land air drilling for example). As a result, there is a need to develop a more nuanced understanding of what constitutes basic training and what is supplemental and to develop programmes accordingly.
- As new technologies are integrated into the workplace, there is a need to ensure that training content and training experiences align with these new developments. Part of this will be to develop simulation technology in line with what people are using on the plant.
- The number of competencies that well-control staff need is increasing and this is placing new demands on education and training within companies. This applies to both new hires and to existing staff who often need additional training.
- The challenge of learning design for a global workforce is significant – how do you cater for different cultural requirements? How do you transfer standards into your organisation to match multiple cultures and departments? The culture within the energy industry is typically very conservative and this can have a negative impact on approaches to training and development.
- Well control sits in the middle of three major conflicts within the industry: operators versus contractors, offshore contractors versus onshore contractors, and US versus Europe. Each conflict has its own dynamic and the need is for all to communicate, collaborate and ultimately align on improving well-control performance.
- All industry participants have a stake in well control: regulators, operators, contractors, well services providers, equipment manufacturers and 3rd party commercial product and service providers. This means that any solution needs to engage effectively with all these participant groups (and navigate the varying interests and priorities of each group).

The primary focus for well control is how to keep hydrocarbons safely in the ground (during the exploration and drilling phase) until production comes along. What are the measures and processes that need to be in place in order to check for uncontrolled hydrocarbons and avoid the type of explosions that led to the Macondo disaster? Well control was the root cause of the Macondo incident. Piper Alpha saw a revolution in personal safety – now Macondo has preceded a revolution in process safety.

The drilling industry has never attempted to address an area of major hazard risk in a comprehensive and collaborative way. However, the current context has created a compelling market need where regulators, operators and contractors all see the value and necessity for a unified industry voice on well control. The WCI is designed to be that unified voice. It is now clear that regulators in every country are demanding that standards are in place and that the industry collaborates to improve well-control safety – if the WCI does not take this on, the regulators will do it themselves and it is better for the industry to develop an effective means of self-regulation rather than having regulators impose their own. The WCI can play a significant role in addressing these challenges.

THE SOLUTION

The decision to establish the WCI was driven by the context outlined above. The principal organisation behind the WCI is the IADC. Following events in the Gulf of Mexico, the IADC senior management recognised the need for a step change in how well-control competencies were developed. At this time, the IADC ran the 'WellCAP' standard and curriculum for rig crews. In recognition of the need to improve and evolve WellCAP, the IADC formed the WellCAP Advisory Panel in 2011 to engage with industry and offer recommendations for a new version of the WellCAP standard and curriculum. The Panel met monthly to discuss its recommendations, and to align these to relevant reports by the OGP.

The other key organisation operating in well-control training standards is the IWCF. Having identified, through engagement with industry, that the demand was for a single standard, governance and administration, the WellCAP Advisory Panel held meetings with IWCF to explore bringing WellCAP and IWCF together. However, talks ended when the organisations could not find an expedient way to come together.

The key recommendation of the IADC WellCAP Advisory Panel was to establish the WCI. The organisation is characterised by the following:

- The governance of the organisation is managed through two bodies: an 18-member Executive Board and a 24-member Advisory Panel, both of which are supported by IADC technical committees.
- The Executive Board consists of CEOs and Senior Vice Presidents from operators, contractors, equipment manufacturers and well service providers. They own the vision and mission and are responsible for the WCI's strategy and tactical goals as well as approving any strategic initiatives the organisation follows.
- The Executive Board is made up of senior representatives from the following firms: Cameron, BP, Maersk Drilling, Shell, Precision Drilling, NOV, Diamond Offshore, Seadrill, Helmerich and Payne, Murphy Oil, Saudi Aramco, Petrobras and Noble.
- The Advisory Panel – which is an independent body – consists of high-level subject matter experts and leaders from across the industry. The Panel focuses

on designing and developing the strategic initiatives set forth by the Executive Board and will provide oversight of WCI operations and performance.

- The WCI will become a global voice for the well-control industry, will promote the WellCAP training standard owned by IADC and will perform an advocacy function for positive well-control initiatives around the world.

The WCI Director of Training and Assessment was brought on board in October, 2013. The Executive Director was brought on in January, 2014. In May, 2014, the Executive Board met for the first time. They defined the mission of WCI as 'improving human performance in well control'.

The initial focus for the WCI has been to assess current well-control training materials and programmes and identify areas for improvement. During the early stages of this process, it became clear that a number of areas could be improved including the following:

- The current model for testing and test development can be unreliable in terms of quality and consistency.
- The focus of training can veer toward pass/fail, rather than knowledge retention.
- The training structure may not be sufficiently well-defined or appropriate.

The way that the WCI addresses these issues is to bring together all sectors of the drilling community to generate recommendations that will influence future iterations of the well-control standard that IADC own. These recommendations build on the work of the OGP Wells Expert Committee and the IADC WellCAP Advisory Panel.

In response to these challenges, the training and assessment standard championed by WCI will focus on learning and learning retention and will provide rig-role-directed learning objectives with an emphasis on kick detection and well shut-in. It will also focus on the use of realistic simulations in a team environment and provide a reliable, secure and trusted testing process. The flexible curricula will cover all well construction disciplines and specialised practices and will offer continuous opportunities for new learning with an enhanced frequency of assessment.

The WCI and IADC have developed curricula at four key levels:

- **An Awareness Curriculum** that is designed for personnel in support roles and identifies a body of knowledge that can be used to provide a basic understanding of well control for drilling operations.
- **An Introductory Curriculum** that is relevant to both surface and subsea operations and that identifies knowledge and a set of job skills that can be used to develop well-control skills for drilling operations.
- **A Driller Curriculum** that builds on the skills introduced in the Introductory Curriculum and that is designed for a number of drilling operation positions including driller and assistant driller.

- **A Supervisor Curriculum** that further develops required well-control skills for drilling operations and that is designed for well-site supervisors, drilling managers and rig managers.

Oil workers clean Pensacola Beach in July 7, 2010 (Photo: Lorraine Kourafas/Shutterstock.com).

A model grading sheet has also been developed for the assessment of learners on drilling simulators (including an assessment designed specifically for supervisors).

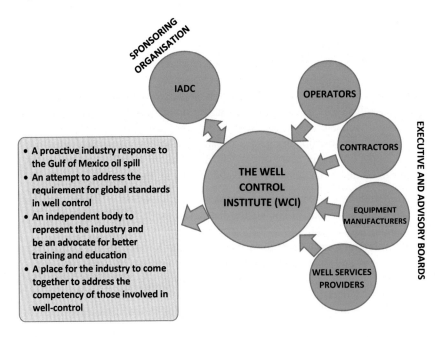

In order to influence the development of the IADC training standard, WCI have produced recommendations that set out the parameters for quality assurance for any training provider looking to offer courses that are based on any of the four curriculum areas outlined above. These recommendations explore the proposed content of courses, the mechanisms and methods of delivery, the facilities and equipment to be used (including information on the use of simulators) and required qualifications for instructors and are designed to offer a comprehensive set of guidelines for any organisation looking to engage in the training and development of well-control staff at any level.

The work of the WCI so far has been characterised by the following:

- The governance structure has been critical. All key participant groups are represented in order to promote the WCI's legitimacy as the voice of, and advocate for, the well-control industry. This also explains why the IADC needed to establish a new organisation as the governance of IADC is dominated by drilling contractors and so would struggle to gain reach across the industry.
- WCI has taken a collaborative approach to building consensus among industry partners. The challenge was to dispel the notion of competition when it comes to well-control performance at the very beginning so that a genuinely collaborative platform could be established to move the industry forward as a whole.
- The governance structure legislates that there is a supermajority requirement to pass any vote. This means that no one group within the Executive Board can dominate the whole through voting blocs, so consensus is required to move anything forward. In the early stages, this put certain constituencies at ease as it meant that they could not be railroaded by others.
- The representatives on the Executive Board are all CEOs or Senior Vice Presidents – none of these organisations passed the responsibility down within the organisation meaning that it has had genuine traction and visibility across the industry.
- The organisation has no particular agenda, either overtly or covertly – they are agnostic of the issues and this is critical to the success they are having. They are a listener and a convener for collaboration rather than a representative of any particular interest.

It should be noted that, at the time of writing, the WCI is very much a work in progress. The organisation has been established to reflect and respond to the challenges and opportunities from across the industry and, as such, is likely to evolve in terms of remit and activities over time.

THE IMPACT

As of 2014, the WCI is still a relatively new organisation although evidence is that the industry is beginning to recognise the value of what is being done. Over time, this will build the strength of the organisation. Their influence is currently felt most acutely in the US and Europe as this is where the leadership within the industry tends to be but there is a recognised need to reach out into other parts of the world and plans to extend into the Middle East, South East Asia, South America, Australia and parts of Africa. In each country, regulators, trade associations and regional operators need to be engaged with on a local and national level as well as continued dialogue with more established international operators.

The success of the WCI can be measured against the following achievements:

- The impact of the new training standard and curricula developed on behalf of IADC alone is broad. By engaging widely and involving the whole industry, the impact is already beginning to be felt worldwide amongst those who develop and deliver well-control training courses. Ultimately, the impact will be felt in the workplace with better trained operators, technicians and managers creating a safer, more productive environment although evidence for this will take time to filter down. It will be possible to better assess success as initiatives take hold over the coming 12–18 months.
- Regulators are beginning to embrace the WCI's role as the industry voice on well-control improvement. They are excited by the structure of the organisation and its remit. They are now keen to see results, and quickly.
- At a minimum, one could say that everyone who delivers, takes and benefits from high quality well-control training has been and will continue to be affected. The intention is that the entire industry is impacted in some way by WCI's work. If WCI's efforts are successful in improving well-control performance, then the industry and ultimately, the public, will benefit as there will be fewer well-control-related incidents worldwide.
- WCI believe that the collaborative, comprehensive approach that they have taken on well control could be replicated and brought to bear on other pressing industry issues. They have provided evidence that a coordinated industry-wide effort can be managed and maintained.
- The WCI is a work in progress. There is much to do and will continue to be for the foreseeable future. However, the organisation believes that the groundwork is being done to sustain this effort and thus make serious improvements in well-control performance.

THE CHALLENGES

Despite evident successes so far, the establishment of WCI has not been achieved without a number of challenges needing to be overcome:

- While it may seem like the industry acts as a whole, in truth there are disparate players with very different and often conflicting needs. Each participant group within WCI's governance structure has its own agenda and bringing those agendas together is no small task. The organisation has so far managed to bring people together by building on common interests and fostering agreement on initiative needs but this is likely to remain as an enduring challenge.

- At the same time as creating the organisation, the WCI has been working on the significant task of improving the WellCAP training standard and curricula for the IADC. Over 150 training providers and 650 instructors are part of the WellCAP network which delivers over 60,000 certificates in well-control training annually. Changing anything about such a programme is an undertaking; changing the fundamentals of that programme is gargantuan. But this has been a critical part of WCI's remit and so had to be done at the same time as creating and promoting the organisation.
- Some participant groups in some parts of the world have not been fully receptive to WCI and its initiatives. Scepticism abounds as to WCI's motives, its governance, its remit and its backing. Addressing this will require ongoing outreach efforts as well as a demonstration of commitment and of the value that the organisation is bringing.
- Managing the expectations of participants was challenging to begin with. Many came to WCI with the belief that this new organisation would be addressing their specific needs, rather than building consensus on what needs the industry has as a whole. This took time to address.
- Balancing efforts with staff resources has been difficult. The organisation runs with only a skeleton staff but is taking on huge initiatives with tight deadlines for performance.
- Regulatory pressures to improve well-control performance preceded WCI's formation, yet now WCI is the target of those pressures. WCI is young, yet is required to mature.
- Part of the challenge for WCI is to understand and communicate what the industry is doing – individual organisations across the industry tend to run their own initiatives and there is not a culture of sharing. In addition, the industry currently has no mechanism for sharing lessons learned. Some kind of infrastructure to support this requirement needs to be established and supported by a change in industry culture.
- Currently, human factors training in well control does not exist in any meaningful way and this has to change (and is something WCI is committed to changing over time).

WCI can be a focal point for the industry through which collective challenges can be addressed. The biggest success factor to making this a reality will be to bring competitive interests together.

THE COST

There has been an initial capital investment on the part of IADC to create WCI and to fund the developments needed for the training standard and curricula improvements.

Once the initial seed funding is no longer needed, the sustaining capital for the organisation is modest (although it is unclear where ongoing revenues will come from).

The Getenergy View

- The WCI is an important – and unique – example of the industry responding to international demand for better standards. It is characterised by an attempt to bring all key partners to the same table – something that has historically been difficult to achieve. The organisation is still in its early stages but it should be congratulated on trying to establish some degree of consensus.
- The establishment of the WCI was a direct response to the Gulf of Mexico oil spill. In many ways, it is revealing that it took a disaster of this scale to get such an initiative off the ground. This also highlights the pressure that the industry as a whole is under to prove that they are serious about safety and that self-regulation can really work.
- One of the key aspects of what the WCI are looking at is human factors. They recognise that human factors are critical to the safe operation of wells. They also recognise that this is an area of education and training that is currently under-served, primarily because there is a lack of understanding across the industry of what human factors really are. It will be interesting to see how WCI are able to bring the issue of human factors to the fore in the coming years.
- The WCI represents a particular response to the challenges of well control. Their establishment also points towards a broader trend – the need to develop mutually understandable standards in the oil and gas industry. The focus on global standards in other areas of the industry is likely to rise up the agenda as a result of the establishment of the WCI.
- The governance structure of the WCI is a recognition of the importance of collaboration across the industry in terms of education and training reform. Everyone needs to be involved in the conversation if progress is to be made (and if recommendations are to be adopted).
- The work of the WCI – and the context within which it operates – demonstrates further the complexity around the ownership and development of standards and their adoption.
- The current level of participation of industry in the WCI should be applauded. That said, future participation needs to take full account of the importance of emergent energy nations as well as those who are well-established.
- The failure of the IADC and the IWCF to effectively collaborate on a single well control standard indicates how far the industry has to travel in understanding and embracing the value of collaboration.

A Note on Sustainability

The WCI is currently a small organisation with low overheads and the backing of the IADC. Short term, the sustainability of the organisation is assured. Longer-term sustainability will be connected to the impact that the organisation can have and the interest it can generate amongst industry partners.

A Note on Replicability

What the WCI is attempting to achieve in well control should be taken note of within other areas of the oil and gas industry. Certainly, the approach to governance, the idea of an organisation that is agnostic of the issues and the concept of a cross-industry champion for standards are all laudable. However, even the alignment to US organisations is a hindrance in the drive to become a genuinely global voice for the industry.

A Note on Impact

It is currently hard to measure the impact that the WCI may have. The organisation is nascent and has a distance to travel before achieving its stated objectives. If the organisation is given the support it needs to coalesce the industry around a single standard, impact could be significant. However, without this support, WCI may lack the necessary tools to effect change and will therefore achieve relatively little.

Glossary of Abbreviations

ACCC	Association of Canadian Community Colleges
AGOCO	Arabian Gulf Oil Company
APEC	Atyrau Petroleum Education Centre
BRIC	Brazil, Russia, India, China
CCPI	Community College Petrochemical Initiative
CMAS	Competency Management System
CoE	Colleges of Excellence (Saudi Arabia)
CTTC	Caspian Technical Training Centre
E&P	Exploration and Production
ENH	Empresa Nacional de Hidrocarbonetos (Mozambique)
FPSO vessel	Floating Production Storage and Offloading vessel
GDP	Gross Domestic Product
GHP	Greater Houston Partnership
HESS	Health, environment, safety and security
HGAs	Host Government Agreements
HISD	Houston Independent School District
IADC	International Association of Drilling Contractors
IHRDC	International Human Resources Development Corporation
ILO	International Labour Organisation
IMF	International Monetary Fund
INEFP	National Institute of Employment and Vocational Training (Mozambique)
INSTEP	Institut Teknologi Petroleum PETRONAS
IOC	International Oil Company
ISD	Independent School District
ISSOW	Integrated Safe System of Work
IWCF	International Well Control Forum
LID	Libyan Integration and Development
LNG	Liquefied Natural Gas
NCS	Norwegian Continental Shelf
NITI	National Industrial Training Institute (Saudi Arabia)
NOC	National Oil Company
NOSS	National Occupational Skill Standards
NVQ	National Vocational Qualification
OAPEC	Organisation of Arab Petroleum Exporting Countries
OECD	Organisation for Economic Cooperation and Development
OGP	International Association of Oil and Gas Producers
OJT	On-the-job Training
OOF	Education Office of Oil-Related Sciences (Norway)
OPITO	Offshore Petroleum Industry Training Organisation

OTP	Operations Training Plant
PAA/VQ-SET	Process Awarding Authority/Vocational Qualifications–Science Engineering Technology
PACE	Prime Atlantic Cegelec Nigeria
PIREP	Programme of vocational education reform (Mozambique)
PPP	Public–Private Partnership(s)
PSA	Production Sharing Agreement
PTS	Petrofac Training Services
PwC	PricewaterhouseCoopers
SAIT	Southern Alberta Institute of Technology
SCPX	South Caucasus Pipeline project
SPSP	Saudi Petroleum Services Polytechnic
SQA	Scottish Qualifications Authority
TVET	Technical and Vocational Education and Training
TVTC	Technical and Vocational Training Corporation (Saudi Arabia)
UEM	Eduardo Mondlane University
USD	United States Dollars
UTSA	University of Texas and San Antonio

Index

Note: Page numbers followed by "b" and "f" indicate boxes, figures respectively.

145